Power and Invention

Edited by

Sandra Buckley

Michael Hardt

Brian Massumi

THEORY OUT OF BOUNDS

...UNCONTAINED

BY

THE

DISCIPLINES,

INSUBORDINATE

PRACTICES OF RESISTANCE

...Inventing,

excessively,

in the between...

PROCESSES

OF

HYBRIDIZATION

Power and Invention

Situating Science

Isabelle Stengers

Foreword by Bruno Latour • *Translated by Paul Bains*

Theory out of Bounds *Volume 10*

University of Minnesota Press

Minneapolis • London

The University of Minnesota Press gratefully acknowledges translation assistance provided
for this book by "Les Empêcheurs de penser en rond" at Synthélabo, Paris.

English translation copyright 1997 by the Regents of the University of Minnesota

The University of Minnesota Press gratefully acknowledges permission to translate and reprint the following essays;
chapter 1 originally appeared as "Complexité: Effet de mode ou probléme?" in *D'une science á l'autre:
Des concepts nomades*, Editions du Seuil, Paris, 1987; chapter 2 originally appeared as "Briser le cercle de la
raison suffisante" in *Cahiers Marxistes*, 168 (1989); chapter 3 originally appeared in a different form as
"Le Réenchantment du monde" in *La nouvelle alliance métamorphose de la science*, Editions Gallimard, Paris, 1979;
chapter 4 originally appeared as "Des tortues jusqu'en bas" in *L'auto-organisation: De la physique au politique*,
Editions du Seuil, Paris, 1983; chapter 5 originally appeared as "Boites noires scientifiques,
boites noires professionnelles" in *La psychanalyse, une science?*, Copyright 1989 Les Belles Lettres, Paris;
chapter 6 originally appeared as "Comment parler des sciences?" in *Chimères*, vol. no. 12 (1991);
chapter 7 originally appeared as "Une science au féminin?" in *Les concepts scientifiques*, Copyright 1989 by
Editions la Découverte, Paris; chapter 8 originally appeared as "Mille et un sexes ou un seul?" in *Les théories scientifiques
ont-elles un sexe?*, Editions d'acadie, Canada, 1991; chapter 10 originally appeared as "Temps et représentation"
in *Culture Technique* 9 (1983); chapter 11 originally appeared as "Drogues: Le défi hollandais" in *Les empêcheurs de
penser en rond*, Synthélabo, Paris, 1992; chapter 12 originally appeared in *L'Autre Journal* 10 (December 1985).

Published by the University of Minnesota Press
111 Third Avenue South, Suite 290
Minneapolis, MN 55401-2520
http://www.upress.umn.edu
Printed in the United States of America on acid-free paper

LIBRARY OF CONGRESS CATALOGING-IN-PUBLICATION DATA
Stengers, Isabelle.
Power and invention : situating science / Isabelle Stengers ;
translated by Paul Bains ; foreword by Bruno Latour.
p.   cm. — (Theory out of bounds ; v. 10)
Includes index.
ISBN 0-8166-2516-6 (alk. paper). — ISBN 0-8166-2517-4 (pbk. :
alk. paper)
1. Science—Philosophy.   2. Inventions—Philosophy.   3. Science—
Social aspects.   I. Title.   II. Series.
Q175.5.S736   1997
501—dc21   97-21760
CIP

The University of Minnesota
is an equal-opportunity educator and employer.

Contents

Foreword

Stengers's Shibboleth

Bruno Latour

Would you say that Isabelle Stengers is the greatest French philosopher of science?

Yes, except she is from Belgium, a country that exists only in part and where, unlike France, the link between science and the state is nil.

Would you say that she is the philosophical right hand of the Nobel Prize winner for chemistry Ilya Prigogine?

Yes, since she wrote several books with him—and yet she has spent the rest of her life trying to escape from the mass of lunatics attracted to this "new alliance" between science and culture that they both created.

Is she a historian of science?

Hard to say. Although she wrote extensively on Galileo, on nineteenth-century thermodynamics, and on chemistry,[1] she remains a philosopher interested in what her physicist and chemist colleagues should understand of their science. Her main object of attention is modern science, and this is what historians and philosophers should study together, no?

You are not going to say that she is an internalist philosopher of science, are you?

Worse than that, Isabelle Stengers is an "hyperinternalist," forcing you always to go further toward a small number of theoretical decisions made by her scientific colleagues. In her eyes, most scientists are often not internalist enough.

But at least don't tell us that she is a Whiggish historian of science looking, like Gaston Bachelard or Georges Canguilhem, for the ways in which hard science finally escapes from history.

She is, I am afraid, much worse. She is "anti-anti-Whiggish," trying to figure out why the anti-Whiggish stance is not the good way to account for what it is to "win" in science, at least not if one aims at convincing the chemists and biologists and physicists she is working with.

But she is a woman philosopher, and at least she must develop some kind of feminist philosophy of science?

There is hardly anyone more critical of the feminist literature, although she uses it extensively and knows it quite well.

Then she must be one of these abstract minds trying to rationally reconstruct the foundations of science and, being busy, erasing all signs of her sex, gender, nationality, and standpoint?

Not at all, there is no one more externalist than her or who reads more extensively in the literature on the social history of science.

What? Does she have any patience for those ridiculous attempts at connecting science and society?

Worse than that, she is addicted to it and knows more "science studies" than anyone else in the field.

Do you mean to say that she likes it because it flatters her radical leanings in politics?

Worse, she wrote on drug legalization, she is a militant in a small left-wing Belgium party, and even went as far as working with charlatans practicing hypnosis and other kinds of unorthodox cures . . . I told you, Isabelle Stengers is always worse! She wrote as much on hypnosis as on physics and she happily compares chemistry laboratories and ethnopsychiatry, going so far as to rehabilitate the word "charlatan."[2]

Then she must be one of these ignorant radicals doing politics because they are unable to grasp the niceties of science?

Not quite, since she does radical politics through the careful definition of what Laplace, Lagrange, Carnot have done with their equations.

I am thoroughly lost . . . then she must be quite a woman?!

Yes, and quite a mind!

But, tell me, how come you have been asked to write a foreword for someone who seems obviously much better endowed in philosophical subtleties, political will, and scientific knowledge than yourself?

This is quite strange, I concur. I guess it is because of the tradition in science studies and in anthropology of the modern world to study "up" instead of "down." Trying to swallow hard sciences had a very good effect on the softer ones. I guess it is the same with Stengers. You grind your teeth on her argument, and you feel much better afterward!

One simple way to define this collection of articles presented in English is to say that they have been written by a philosopher interested in the very classical question of distinguishing good science from bad. Her new solution to this old problem will be, however, difficult to grasp both for science studies and for philosophers, and it is a solution that requires some clarification. Isabelle Stengers does not share the antinormative stance of most recent historians and sociologists of science and has no qualms about looking for a shibboleth that will help sort out science from nonscience. In this sense, but in this sense only, her work is marginally more acceptable to Anglo-American epistemologists than those of "science studies" who shy away from any normative position. Philosophers will at least be able to recognize that here is someone who is not complacent vis-à-vis the production of bad science and who shares their will for a good cleansing job. The difference, because fortunately there is one, lies in the fact that her own touchstone means getting rid of most epistemologists and quite a lot of hard sciences! So the normative goal is similar but the principles of choice are radically different.

Where does this difference come from? Isabelle Stengers has chosen to look for a touchstone distinguishing good science from bad not in epistemology but in ontology, not in the word but in the world. This is the trait that no doubt makes her work sound so bizarre to the innumerable descendants of Kant and Wittgenstein, people in the ranks of philosophy and social construction alike. The only way to determine why a statement can be accurate or inaccurate has been, since at least Kant's "Copernican revolution," to look at how the mind, the language, the brain function. While we are disputing among humans how to have a faithful representation of the world, the world itself, in the meantime, remains completely out of the scene, serenely and obstinately similar to what it is. Amusingly enough, this presupposition is shared by the very classical philosophers who insist on radically separating epistemological from ontological questions and by the radical sociologists who insist very classically on leaving the world outside of our representations. Everyone seems to agree that in sorting out good science from bad, only the human side has to be interrogated, not what the things do since they cannot be the source of our misinterpretation about them, nor of our consensus on what they finally are. Stengers's solution to the question of how we come to agree or disagree about the world is completely non-Kantian. To be sure, our society, language, mind, and brain could be cause for some misunderstanding, but the main partner to be interrogated for sources of uncertainties is the complexity of the world, which does not wait outside and does not remain equal to itself. Against epistemology and against social construction, Stengers directs our attention to the ways in which the world is agitating itself and puzzling us.

This is especially clear in her "first period," so to speak, which makes up Part I of this book, during which she cooperated with Prigogine to understand until which intimate level the chaotic agitation of the world itself could modify our definition of science. Here is the first dramatic case of a normative touchstone to sort out good science from bad that does not look for the limits of human representation but for the world's ways of marking the limits. For the authors of those books (the success of which has been phenomenal, in French at least),[3] any discipline that does not take into account the arrow of time (or better the arrows of times) is not a science, no matter how hard, respectable, or highly objective it looks. This was the first application of what could be called an "ontological touchstone" that is clearly different from the one used by epistemologists since it throws into the dustbin the very disciplines that had been used until then as the standard to judge all other efforts at scientificity. Clearly, epistemology, with its attention to language, representation, clarity, and rigor, is not equipped to sort out good science from bad, since it has been unable to detect that time, irreversibility, complexity, and agitation have been papered over by most authors in classical physics. "You blind guides! You strain out a gnat, but swallow a camel" (Matt. 23:24).

If Isabelle Stengers had stuck to this first definition of her ontological touchstone, she would have remained a commentator of Prigogine's fights and quarrels with his peers and other "dear colleagues." She would have been the philosophical henchman of a highly controversial chemist. For the same reason, she would have remained the deep admirer of Stephen Jay Gould and of all the evolutionary theorists engaged in a constant demarcation between good and bad narratives about a subject evolution that has many more degrees of freedom than our representations of it. But in this way she would have forever stayed a classic philosopher of science, tempted by the rather romantic idea of a science of time reconciled with the rest of culture. But, as she discovered (in part because of the success of this first work), the "new alliance" between science and culture cannot be so quick and cheap. No matter how time-dependent a science of phenomena far from equilibrium can be, it remains a science, that is, an attempt at stabilizing the world. But what is a science? This "second period" corresponds, for Isabelle Stengers, to a series of articles and books written in her own name in which she explores another version of an agitated world, the version offered by the readings of Alfred North Whitehead and Gilles Deleuze.[4] In this second layer, so to speak, of the same ontological touchstone, she moves from philosophy of science to philosophy proper, from the question of an agitated chaotic world of science to the ontology of a world that is itself the main cause of most of our uncertainties.

In countries where philosophy has been separated into epistemology on the one hand, and history of ideas on the other, it is very hard to locate a philosopher like Stengers who takes up the normative task of epistemology but who carries it out by using the tools of metaphysicians like Leibniz or Whitehead, who for generations have been taught (or not taught at all) as so many dead white males. For her, metaphysics is epistemology pursued by other means, a serious task that requires the collective wisdom of the whole history of science and thoughts and that cannot disdain any of the rejected claims of past philosophy or underdog sciences. As will be clear in reading this volume, the effects of this writing strategy are very strange, especially when famous scientists — Galileo, Einstein, Poincaré, Planck — are read not as those who broke away from philosophy but as those who can be elevated to the level of great and controversial metaphysicians, fighting as equals with obscure figures of medieval theology like the unexpected Étienne Tempier, a favorite figure of Stengers's bestiary. The effect will be even stranger when read by a historian, who will not understand how one can jump so easily through centuries, and yet who will have to recognize that at every point Stengers's accounts, if not historical in character, are at least "history compatible." No internalist philosopher has provided more hookups in her argument to plug in the most advanced "peripherics," as one says in computer parlance, of social history of science. The result is a prose that is not always easy to follow, but in which science and philosophy are forced to again become hard ontological and political questions — a very strange mixture for which Stengers has since found a new beautiful name, a name taken out of Kant's very heirloom, that of *cosmopolitics*.[5] As one of her many students said in jest, the new question is no longer to decide if a statement is PC, but if it is CC, meaning "cosmopolitically correct."

What is a CC statement? What is a statement that pursues the task of demarcation all the way to ontology? One thing is sure: if the reader applies to Stengers the traditional settlement between science, politics, ethics, and theology that characterizes the modernist idiom, then her attempt will be hard to follow and so will Whitehead's and Deleuze's. All these authors do not recognize the settlement that can be defined in the following way: first, a world outside untouched by human hands and impervious to human history; second, a mind isolated inside its own mind striving to gain an access to an absolute certainty about the laws of the world outside; third, a political world down there, clearly distinct from the world outside and from the mind inside, which is agitated by fads and passions, flares of violence and eruptions of desires, collective phenomena that can be quieted down only by bringing in the universal laws of science, in the same way that a fire can be extinguished only by water, foam, and sand thrown from above; and fourth, a sort of position "up there" that serves as a warrant for the clear separation of the three spheres above, a view from nowhere that is occupied either by the God of ancient religions or in recent times by a more reliable and watchful figure, that of the physicist-God who took upon himself—it is definitely a he!—to make sure that there are always enough laws of physics to stop humans from behaving irrationally. No progress can be made in the philosophy of science if the whole settlement is not discussed at once in all its components: ontology, epistemology, ethics, politics, and theology. This point of method has been made clear, by the way, through the so-called science wars that bring all the distinct threads of the old settlement together again—except that, as usual, history repeats itself as parody...

It would be an understatement to say that Stengers is not a partisan of that sort of constitution. But neither is her position that of the critical stand taken by social constructionists who prefer to say that the connections between these four spheres do exist but are unfortunately detached; that the inside mind does not have a safe connection with the outside world, which means that no indisputable laws of science can be brought to quench the political unrest of the unruly masses, which has the consequence that any godlike figure will remain forever totally impotent. But, like her only true mentor Deleuze, Isabelle Stengers has no patience for critical thinking. She does not ever say that those spheres are necessary and that connections between them have been, alas, severed. She claims that those spheres do not exist at all and have never existed: the world is not outside, the mind is not inside, politics is not down there, and as for the physicist-God, he possesses no view from anywhere because there is no longer any need for this sort of arbitration work. Anglo-American readers often have difficulty in accepting that one can think and write out of the Kantian settlement altogether and thus also out of its critical appraisal.[6] If it is not Kant, they assume, then it has to be Wittgenstein. If the modernist foundation is impossible, then it has to be the constant irony exerted against the lack of any foundation.

Isabelle Stengers does not like irony more than denunciation. She proposed once to define philosophy as "l'humour de la vérité."[7] Like most philosophers of her tradition, she lives in a world of events, not in a prison of words trying desperately to represent an absent and faraway state of affairs. Propositions, to take up one of Whitehead's key words, are moving through and are not human interpretations of things-in-themselves that would be out there remaining indifferent to our fate. Politics is not about quieting down passions and emotions by bringing in rationality from above, but about deciding, on the spot, what is the good proposition that does justice to an event. The mind is not an isolated language-bearer placed in the impossible double bind of having to find absolute truth while it has been cut off from all the connections that would have allowed it to be relatively sure—and not absolutely certain—of its many relations. It is a body, an ethological body, or, to use Deleuze's expression, a "habit of thought." The country in which those noncritical philosophers travel is totally different from the lunar landscape in which epistemologists and social constructionists have been waging their two-hundred-year war. One is not the critique of the other. They differ like nonmodernity from modernity, like the surface of the green planet differs from that of the moon.

If there is no separation between world and word, between propositions and substances, between what happens to humans and what happens to nonhumans, then Isabelle Stengers should fall, one will object, either into the physicalism (or organicism) of which Deleuze has been so often accused, or else into the generalized Machiavellianism of some sociologists of science. She might have escaped from the modernist settlement, but, one could object, she has to fall into the double peril of "everything is nature" or "everything is politics." The trap of misunderstanding her is ready, wide open, and well oiled. This would be to forget that she is, at heart, a normative philosopher sticking firmly to the classical task of distinguishing good from bad science. Thus, there is a distinction at work that saves her from all sorts of monisms, including Deleuze's. It could be called "risky construction" because it is a specific type of construction that takes risk as its touchstone. Since this is the point that will be hardest to grasp in transferring Stengers from French to English, it is better to take another close look at the ways in which her shibboleth strikes through the sciences in the most unexpected ways.

Let us remember that the distinction she tries to make is not the one between true and false statements, but between well-constructed and badly constructed propositions. A proposition, contrary to a statement, includes the world in a certain state and could be called an event, to use another key Deleuzian concept. Thus a construction is not a representation from the mind or from the society about a thing, an object, a matter of fact, but the engagement of a certain type of world in a certain type of collective. Constructivism, for Stengers, is not a word that would have an antonym. It is not, for instance, the opposite of realism. Thus, constructivism is the opposite of a pair of positions: the twin ones obtained after the bifurcation, as Whitehead says, between world and word. In this way, "social construction" is not a *branch* of constructivism, but the *denegation* of any construction, a denegation as thorough as that of realist philosophers. So we do not have to choose between realism and social construction because we should try to imagine some sort of mix-up between the two ill-fated positions. Rather, we have to decide between two philosophies: one in which construction and reality are opposite, and another in which constructing and realizing are synonymous.

This is why, to make her point clearer, Stengers adds to the notion of construction that of risk. There are constructions where neither the world nor the word, neither the cosmos nor the scientists take any risk.[8] These are badly constructed propositions and should be weeded out of science and society; that is, they are not CC (cosmopolitically correct), no matter how PC they appear. On the other hand, there exist propositions where the world and the scientists are both at risk. Those are well constructed, that is, reality constructing, reality making, and they should be included in science and society; that is, they are CC, no matter how politically incorrect they may appear to be. In Isabelle Stengers's hand, this risky constructivism is an extremely powerful demarcation criterion because it strikes at first unexpectedly, making her (as well as her readers) take a lot of risks—I myself bear many scars because time and again I have been too clumsy to predict where the shibboleth will break up my halfhearted arguments. "Women, fire, and dangerous things" is an expression that would nicely fit Stengers's handling of her many friends.[9]

As will be clear in reading the articles in this book, many examples are offered of this dangerous trial by fire. When she was working with Prigogine, Stengers used her ontological touchstone to render unscientific all the disciplines that designated that time could be one of the essential features "proposed" to them. Through typically Stengersian unexpected connections, she then began to work with Léon Chertok, a fascinating psychiatrist living in Paris who used hypnosis in his therapy despite the discontinuation of that technique by Freud and his many disciples.[10] Nothing more clearly demonstrates the originality of her touchstone than comparing the effects of that same principle on physics and on psychiatry. One could expect that the application of any demarcation criterion will rule psychoanalysis out of science. This is, after all, what Karl Popper had done. Maybe—but then it should certainly rule hypnosis out as well! What could be said of a criterion that kicks reversible physics and scientistic psychoanalysis out of science and that keeps time-dependent chemistry, chaotic physics, and also hypnosis on the right side of the border? There must be a deep flaw in Stengers's demarcation criterion.

This, however, is not the case if we follow how what I call her risky construction works.[11] The principle of sufficient reason cannot be sufficient if it keeps reversibility in and leaves irreversibility out of the picture. Freudian psychoanalysis tries to imitate the causal principle of sufficient reason by behaving like a science—but a science conceived in the way that has not yet applied the Stengersian principle! Freud, terrified by the incredibly fast and complacent reaction of his patients to his influence through hypnosis, decided to imitate "hard science" and make sure that he was not the cause of behaviors in his patients. He protected himself by the famous phenomenon of transference and, from then on, dealt with patients through a purified analysis that tried to apply quasi-chemical procedures to the "laboratory" of the couch. What is the result of this Freudian "will to science"?[12] The elimination of influence from psychiatry in the same way that the arrow of time has been eliminated from physics.[13] Yes, psychoanalysis is a science, but that in itself is not enough of a guarantee to be kept on board since reversible physics has been thrown out for being exactly as badly constructed! The same principle strikes twice with the opposite result: one should not eliminate from a discipline what constitutes its main source of uncertainties and risk, reversible time in the case of nonhuman phenomena, susceptibility to influence in the case of human phenomena.

The paradoxical result of the work of Stengers with Chertok is the same as that of her earlier work with Prigogine: the question is not to decide what is scientific and what is not, that is, to demarcate science from nonscience, but to distinguish within the sciences, or better, within the cosmopolitics, the procedures through which the scientists expect to run as much risk as their subjects. Paradoxically, it is not because it has foolishly tried to treat humans like nonhumans that psychoanalysis fails the Stengersian test. It is exactly for the opposite reason: psychoanalysis fails because it treats humans like no hard scientists would dare treating their objects, that is, without giving them a chance to redefine, on their own terms, what it is to be interrogated by science. There is, of course, a difference, but it is not the ancient one that distinguished an objective matter of fact that can easily be mastered from a human soul that would resist any attempt at mastery.[14] The difference comes from the innate resistance of nonhumans to be taken up by science—the irreversibility of time remaining, for Stengers, the paradigmatic example—whereas humans are incredibly complacent, behaving too easily as if they had been mastered by the scientist's aims and goals. This was Freud's real "scientific" discovery, which, unfortunately, he failed to see because of his wrongheaded idea of what was a science. In

Stengers's trial, the presence or absence of the trappings of science proves nothing at all. If one is daring enough to take the test, one should be ready to demonstrate instead that the questions raised by one's experiment are at risk of being redefined by the phenomena mobilized by the laboratory or by the theory.

If it looks superficially reminiscent of Popper's falsification criterion, one has only to see which sciences it throws out and which one it keeps on board to measure the complete difference between the two epistemologies. Popper's touchstone is as good as a white coat. It is easy to don but it does not make a scientist out of the one who wears it. On the other hand, Stengers's criterion sees the sheep through the wolf's fur coat! Evolutionary theory, that of Gould's *Wonderful Life*, for instance,[15] is kept after Stengers's trial because it fits exactly her requirements: every species forces the natural historian to take as much risk to account for its evolution through an innovative form of narration as it took the species to survive. However, Popper's razor excises Darwinism out of science, together with Marxism, history, and Freudianism, on the very slim pretext that it cannot be put to the test. But which test? The one where scientists master all the inputs and outputs and leave the objects no other freedom than the ability to say "yea" and "nay"! It is a very poor science in which things have no more to say than the white and black pawns in a game of Master Mind and where the wild imagination of the scientist does all the rest of the talking.

Fortunately for science, there are endless situations in which scientists can be left voiceless by the wild imagination of things proposing to them what to say. Amusingly enough, falsification misses these situations as well and honors fake imitations with the Medal of Science. For instance, Popper's criterion will keep Stanley Milgram's impeccably falsified experiments since it puts to the test the wild hypothesis of an innate obedience to authority among American students.[16] What could be more scientific than this most famous experiment in psychology? Has Milgram not all the controls required? Has it not included all the blind tests? His experiments, however, are torn apart when Stengers's criterion is applied, since it ideally fits her case of a bad construction where nothing can be learned from the students subjected to Milgram's power, and where Milgram does not even learn the only lesson he could draw from this disastrous experiment: that he is the only torturer in town whose mad power over subjects should be interrogated. Oh, yes, there is a blind test indeed, but the blind is not the one you would think! This lesson can be drawn on an indefinite number of experiments published in excellent journals, all equipped with referees and fact finders.

The biggest difference between Popper's and Stengers's criteria resides somewhere else, though. Popper's falsification implies a complete power of the scientists themselves to sort out their own inventions. It was made for that, to protect scientists against any encroachment from society. On the other hand, Stengers's shibboleth allows her to look everywhere for the conditions where power is counterbalanced by the invention of those who are talked about. Most of the time this means, of course, getting out of science. This exit from the classical questions of epistemology will be made clear in the last part of this book. Popper remains a traditional philosopher of science, and if so many disciplines fail in his eyes, it is because they look suspiciously close to the horrible quandaries of political life where nothing can be safely falsified, human masses having great difficulty, it seems, in limiting their presence in the forum to "yea" and "nay." If society has so many enemies, it is because, first of all, society is the enemy of science. Popper's philosophy of science might have been well adjusted to the political task of the 1930s, but Stengers's aims at understanding how we should live now within and without the limits of science proper — or should I say popper? This means that one has to leave the confines of science to see how the same risky construction could be applied to collective situations that have none of the features of scientific facts and where, nonetheless, the same dilemma can be observed.

How does this second task work? Remember that Isabelle Stengers is uninfluenced by how much resemblance a practice has with science conceived in a Popperian way. She knows from the inside — and this is why it is so important for her to be a thorough internalist — that the question is not so much how you can mimic a science (it has become so easy, so safe, so cheap that even sociologists can do it!) but how much risk one can take in allowing one's words to be modified by the world. This means that a practice which, on the face of it, looks completely unscientific, such as hypnosis, drug addiction, ecological politics, ethnopsychiatry, or AIDS patients groups, may manifest features that render them closer to some of the most abstract and daring hard sciences. This is where Isabelle Stengers will appear the most controversial and probably the most original in the American context of higher superstitions and lowly politics. A difficulty arises for me at this juncture, however. The articles assembled here are not doing full justice to the importance of this second aspect of her cosmopolitics. Like many of the political insights of Deleuze and Guattari, they remain largely influenced by a conception of left-wing radicalism that has not yet been renewed as forcefully as science has been. This is a case, so to speak, of an "unequal development" of a theory. In the last part of this book, Stengers has not done for power, society, and domination the job of redemarcation that she has done to hard science in the earlier sections. She relies too heavily on the tools of social history that can be taken off the shelf.

But the direction in which she has been heading since these articles were written seems to me clear enough and fully vindicates her intention of practicing a genuine cosmopolitics that will strike the knowledge-talk and the power-talk as well. It is thus to her more recent work that one has to turn to show the full import of her shibboleth. In going from the laboratory of chemistry to the platforms of politics, Isabelle Stengers does not try to decide which science is politically correct, which politics is ideologically sound. She does not flee to society because she had been disappointed by science as do so many critiques of Western "thanatocracy." Radicals should be ready to be more fiercely sorted out on their planks by the same demarcation criterion as white-coat physicists at the bench. Stengers does not appeal from the limits of an objective science to the passion of radical politics; she ruthlessly sorts out the objects of scientists and the passion of militants. In every case, and at every juncture, she remains unimpressed by domination, no matter if it comes from the ranks of science or from the ranks of social powers. In both, she looks for the sources of invention that have been missed.

If scientists are surprised by the ways she demarcates good science from bad, the many people who, from the ranks of feminism, ecology, and leftism, think she is their ally should brace themselves for some hard lessons, more exactly, from the lessons she keeps drawing from hard sciences. Going from science to politics is not, for her, going from stringent constraints to more relaxed ones, but keeping exactly the same objectives with a total indifference to what is science and what is society. Domination in politics has many of the same ingredients it has in the laboratory, that is, the inability to allow the people one deals with any chance to redefine the situation in their own terms. If this principle subverts many disciplines from the inside, it subverts even more political stands from the outside, and especially so many of the "standpoint politics" where the outcome of the analysis is entirely determined from the start by the position of the speaker.[17] If Milgram is taken as the emblematic bad experimenter, not giving the students he is torturing a chance to become torturers, what should be said of those thousands of radical tracts where the things to be studied—science, art, institutions, medicine—have no chance to say anything other than that they have been marked by the domination of white male capitalists? Like most critical thinking, they reproduce exactly at the outcome that was expected from the beginning, and if they have to be rejected, it is not because they are political, and not because they are not scientific enough, but simply because the writer incurred no risk in being kicked out of his or her standpoint in writing them. The application of Stengers's criterion on "cultural studies" remains to be seen, but it will be even more entertaining than what it did at the bench. The equation is simple, although very hard to carry out: no risk, no good construction, no invention, thus no good science and no good politics either. Such is the first plank of a party that does not have many members yet!

Stengers's request to be cosmopolitically correct cuts both ways, and cuts hard. In the obscure fights of the science wars, one can safely predict, she will be seen as a traitor to all the camps, not because she is "in the middle"—no one is less of a middlewoman than her, no one is less an adept of the golden mean!— but because she imposes on all protagonists a criterion that they will do their utmost to escape. Although this book appears in a series called "Theories Out of Bounds," no theory is more binding than Stengers's new demarcation criterion. Having often tried to escape its binding strength only to find myself forced to use it again, it is a great pleasure (and I say it with some glee) to imagine that English-speaking readers are now to be enmeshed in this most daring enterprise that we, in the French-reading world, had to take into account for so long. It is my hope that they will learn more than I did (this is unlikely) in those twenty years when I tried to profit from her marvelous "habits of thoughts," and also my hope that they will be forced even more than I was (this is more unlikely) to modify their definition of hard science and of radical politics by using Stengers's shibboleth and pushing it everywhere—even against her, if needs be!

Foreword

1. Bernadette Bensaude-Vincent and Isabelle Stengers, *A History of Chemistry* (Cambridge: Harvard University Press, 1996).

2. Tobie Nathan and Isabelle Stengers, *Médecins et sorciers* (Paris: Les Empêcheurs de penser en rond, 1995).

3. Ilya Prigogine and Isabelle Stengers, *La nouvelle alliance, métamorphose de la science* (Paris and New York: Gallimard/Bantam, 1979); Ilya Prigogine and Isabelle Stengers, *Entre le temps et l'éternité* (Paris: Fayard, 1988).

4. Isabelle Stengers, *L'invention des sciences modernes* (Paris: La Découverte, 1993); Isabelle Stengers, ed., *L'effet Whitehead* (Paris: Vrin, 1994).

5. Isabelle Stengers, *Cosmopolitiques*, vol. 1, *La guerre des sciences* (Paris: La Découverte and Les Empêcheurs de penser en rond, 1996); vol. 2, *L'invention de la* *mécanique: pouvoir et raison* (Paris: La Découverte and Les Empêcheurs de penser en rond, 1996).

6. They have the same difficulty with another of Stengers's earlier fellow travelers; see Michel Serres, *Conversations on Science, Culture, and Time with Bruno Latour* (Ann Arbor: University of Michigan Press, 1995).

7. The "humor of truth" instead of "l'amour de la vérité," the love of truth. The pun unfortunately does not work in English.

8. For another attempt at bridging the gap between "science studies" and normative questions, see the notion of "epistemic virtue" developed by Adrian Cussins in "Content, Embodiment and Objectivity: The Theory of Cognitive Trails," *Mind*, vol. 101, no. 404 (1992): 651–88.

9. George Lakoff, *Women, Fire, and Dangerous Things: What Categories Reveal about the Mind* (Chicago: University of Chicago Press, 1987).

10. Léon Chertok, *L'hypnose, blessure narcissique* (Paris: Laboratoires Delagrange, 1990); Léon Chertok and Isabelle Stengers, *Le cœur et la raison. L'hypnose en question de Lavoisier à Lacan* (Paris: Payot, 1989); Léon Chertok, Isabelle Stengers, and Didier Gille, *Mémoires d'un hérétique* (Paris: La Découverte, 1990).

11. For another application of this principle on ethology, see the enterprise of another Belgian philosopher, a colleague of Stengers, Vinciane Despret, *Naissance d'une théorie éthologique* (Paris: Les Empêcheurs de penser en rond, 1996).

12. Isabelle Stengers, *La volonté de faire science* (Paris: Les Empêcheurs de penser en rond, 1992).

13. It is precisely this "influence" that Tobie Nathan, a student of Georges Devereux and a close associate of Stengers, has been reintroducing into very risky clinical procedures; see Tobie Nathan, *L'influence qui guérit* (Paris: Éditions Odile Jacob, 1994).

14. This is the source of another misunderstanding, this time with the tenants of the hermeneutic tradition who believe that because philosophers like Stengers, Deleuze, or Serres are attacking scientism, they bring water to their mill and will help them defend the subject against the tyranny of the object. Quite the opposite. The subject has to be treated, they propose, at least as well as the object! It is the object of science who does the job in the hermeneutic circle, not the human subject always ready to imitate a machine, or what he or she imagines the machine to be.

15. Stephen Jay Gould, *Wonderful Life: The Burgess Shale and the Nature of History* (New York: W. W. Norton, 1989).

16. Stanley Milgram, *Obedience to Authority: An Experimental View* (New York: Harper Torchbooks, 1974).

17. In this way, she is even further from Michel Foucault's famous knowledge/power than from traditional epistemology, and this offers still another source of possible misreading in an Anglo-American context strongly influenced by a "socialized" version of Foucault.

Science and Complexity

O N E

Complexity: A Fad?

FOR SOME years, the theme of complexity has played an ambiguous role in discourses on science. It allows one both to defend science against the charge of "reductionism" and at the same time to envisage science's conquest of what until now had escaped it. For example, what is meant by the statement: "The brain is complex?" This expression can just as easily figure in an attack against those who seek to explain the brain by the simple elements of which it is composed, the neurons, as in an introduction to the notion of complexity. In the first case, the word "complexity" is used to ward off an operation of capture; in the second, it announces its possibility.

The "discovery of complexity" is often spoken of as a distinguishing feature of the current era. On the other hand, as soon as the term "complexity" is employed, one hears jokes and pejorative remarks about the notion, which is dismissed as a media invention having no real meaning. These two types of reactions, which tend to reinforce one another, indicate from the outset the nature of our problem.

A first point to emphasize: complexity, whether denounced as a diversion or announced as science's redemption, belongs to a discourse *about* science. Admittedly, the adjective "complex" has appeared in a certain number of disciplines, particularly in mathematics and physics. And those who talk about "the discovery of complexity" often refer to one or another of these disciplines. However, none of these references is, in itself, sufficient to justify the theses that they illustrate. When physicists constructed the formalism of quantum mechanics, it could be said that a problem was *imposed* on them, whether they had foreseen it, searched for it, or simply accepted it. The situation is much less clear regarding "complexity." One cannot, here, designate the problematic node that would "force" those who speak of it to recognize a situation as "complex," let alone draw far-reaching conclusions on that basis.

In other words, the putative "discovery of complexity" designates something altogether different to the type of episode that, in collective memory, punctuates the history of certain sciences. The usual progression implies a passage from a state of ignorance to a state of knowledge, or the sudden appearance of an unexpected problem that completely disrupts the anticipations of a science. But if the possibility that complexity has been "discovered" is to be taken seriously, the issue is not the imposition of a question but the *growing awareness of a problem*, an "awareness" that may have been provoked, but was not imposed, and thus can easily be denied by those who fail to see the point of it.

If it is indeed fair to say that there has been a growing awareness of a problem, its context is singularly ambiguous. In fact, the first thing that might strike someone interested in the discourse on complexity is the revival of a kind of classical scientism. What seems to happen is that themes of world crisis, and a questioning of the presuppositions that allowed us to underestimate the crisis or to think of it as epiphenomenal, are interwoven with the themes of a "new rationality." This is an eminently classical scientism, in that the renewal of the scientific knowledge that was initially critiqued is heralded as a *solution* to ethicopolitical problems.

One might also take as a case in point the grand cosmic-social "frescoes" that depict the progressive complexification of the world, from the big bang to the problems of contemporary society. Contrary to the representations of nineteenth-century positivism, a book like Hubert Reeves's *Atoms of Silence: An Exploration of Cosmic Evolution*[1] no longer emphasizes reassuringly linear progress: with complexification comes instability, crisis, differentiation, catastrophes, and impasses. But the fresco of cosmic complexification is nevertheless reassuring, in the same way that positivist representations were, in that it presents itself as a theory, in that it inscribes the overarching questions of our era in a narrative frame that connects them to crises of matter and the genesis of life.

This way of using the notion of complexity or complexification might in itself provide sufficient grounds to condemn the notions; for if, a priori, the discourse on complexity has meaning, that meaning cannot be homogeneous to the science it critiques. The vision of a complex world per se cannot be substituted for another scientific vision of the world; it is the notion of a vision of the world, from the point of view of which a general and unifying discourse can be held, that in one way or another must be called into question. Failing that, it is certainly possible to comment on the interest of new types of formalization, on new physical and mathematical objects, on new methods of description that have been described as "complex"; it is also possible to foresee that these new modes of questioning will have effects in other domains of knowledge. But it is not possible to speak of a "discovery of complexity," in the sense that this would translate not only into an enlargement but into a transformation of the field of scientific knowledges.

Under such conditions, to be interested in the question of complexity is not a neutral activity. The least risky approach would certainly be to limit oneself to an attitude of denunciation; as we have seen, this approach has solid arguments in its favor. It is easy to raise the suspicion, and even to demonstrate, that, behind the good intentions and generous declarations, the relations between different knowledges remain hierarchical, loaded with ignorance and even contempt. Easy critique, however, has always had a certain sterility: namely, its eternal confirmation of the possibility and necessity of maintaining an attitude of reserve and irony in the face of scientific arrogance. The issue we will raise here is more difficult and risky. Can we attribute a general applicability to the theme of complexity without authorizing generalizing pretensions? Can we use what present themselves today as "complex objects" to underline the general problems they raise, rather than the particular models of solution they determine?

Relevance and Risk

If the notion of complexity must have a general bearing on the theories and/or practices of the contemporary sciences, it needs to have a meaning that is not from the outset dependent on a particular discipline.

It seems to me that if the theme of complexity is potentially *interesting*, and perhaps worthy of surviving compromizing usage, it is not so much because it delineates the characteristics of a "new science" that we were previously unable to imagine, but rather because it rekindles and highlights what is without doubt the most genuinely original aspect of what is called "modern science." As Jean-Marc Lévy-Leblond reminds us, the function of scientific thought has less to do with its "truth" than with its *astringent effects*, the way it *stops thought from just turning in self-satisfying circles.*[2] The theme of complexity allows the uncoupling of two dimensions that are often inextricably associated in discourses for or against the sciences: the power of the analytical approach and the peremptory judgments that it appears to authorize; and "scientific rationality" and the fearless production of "scientific views of the world."

I think that this uncoupling sharpens the sense of a term that is at the heart of scientific practices, but that tends to disappear in public discourses on science: that of *relevance*. What is noteworthy about "relevance" is that it designates a relational problem. One speaks of a relevant question when it stops thought from turning in circles and concentrates the attention on the singularity of an object or situation. Although relevance is central to the effective practices of the experimental sciences, in their public version it often boils down to objective truth or arbitrary decision: to objective truth when the question is justified by the object in itself, and to arbitrary decision when it refers to the use of an instrument or experimental apparatus whose choice is not otherwise commented on. In the first case, the response appears to be "dictated" by reality. In the second, it appears to be imposed by the all-powerful categories of which the investigative instrument is bearer. Relevance designates, on the contrary, a subject that is neither absent nor all-powerful.

What are the "right questions"? What is the "relevant" point of view? This is the fundamental question of experimental science.

Let's take the most classic example: Galileo's theory of falling bodies. This theory entailed a wager on reality in the sense that it determined a priori what, in the observable fall, would be considered significant and what should be judged an insignificant perturbation. In this case, the theory implies a conceptual separation between the fall as it would occur in a vacuum and aspects that are tied to air friction. This very separation constitutes a theoretical decision—which the Aristotelians, it should be noted, judged inadmissible. In Aristotelian physics, falling, as a concrete movement, implied air. This theoretical decision is at the same time a practical one. It guides the process of preparation and experimentation: the "phenomenon" is technically redefined "in the laboratory" and purified to the extent possible of everything assimilable to noise. It is at this point that the wager pays off, or is lost. It pays off if the predictions that the theory allows to be constructed are coherent with experiments on the purified system conducted in accordance with the theoretical criteria.

Experimentation in this context, is a *risky* process. It assumes that the phenomenon as isolated and reworked under laboratory conditions is essentially *the same* as the one found in "nature." The notion of "scientific view of the world" tends to underestimate this risk, presupposing that a question relevant in certain experimental conditions will remain relevant, and thus can serve as a model for the generalization of that particular worldview's corresponding mode of distinguishing between what is significant and what is insignificant.

It is in this framework that the question of complexity takes on precise meaning. This question comes to the fore when the relevance of a simple model becomes an issue, along with the relevance of the prolongation of the process authorized by the model to phenomena subsequently judged "complicated" but not intrinsically different. Thus, the question of relevance and that of reductionism are connected. The reductionist operation is, by definition, an operation of prolongation. It is marked by two kinds of affirmation: "such a phenomenon *is only* ..." (this means that the simple model is able to define the questions relevant to its subject); "if we were more precise observers, if we had more knowledge, more methods of calculation, more facts, we could ..." These two affirmations are sometimes completely justified. At other times, they presuppose a passage to the limit, the critical nature of which is indicated by the question of complexity.

It should be emphasized that the question of complexity as I have elaborated it is truly a product of the analytical spirit. Analysis and reductionism are too often lumped together in the same critique. But, as we will see, it is quite possible for the analytical method to directly contradict the generalization of reductionism. Far from entailing the idea of a more simple world, analysis can lead to the conclusion that we do not know what a being is capable of. One way or another, reductionism always ends up " ... is only ... "; the analytical method, on the other hand, may lead to "this ... , but in other circumstances that ... or yet again that. ... "

Let's take an example. One speaks of complexity with respect to "strange," "chaotic,"[3] or "fractal" attractors. An attractor is a stationary state or regime toward which an evolution described by a well-determined system of equations leads. Usually, an attractor is stable: different sets of different initial conditions determine an evolution toward the same attractor (for example, a state of thermodynamic equilibrium, the immobile state of a real pendulum, from which one has not abstracted friction; or a "limit cycle"). Once this attractor has been reached, the system will no longer spontaneously depart from it, fluctuations aside. "Strange attractors," on the other hand, do not have this property of stability. Two neighboring initial conditions can generate very different evolutions. The slightest perturbation can push the system from one regime into a very different one. Instead of stabilizing into a predictable and well-determined state, the system wanders between possibilities; in other words, although governed by deterministic equations, it adopts an aleatory behavior.

The possibility of representing an observable phenomenon through deterministic equations that link its different determinations in a coherent manner is usually assimilated to a sort of terminus ad quem. But when a system is identified as having a strange attractor, the equations that represent the system do not erase the uncertainty of its behavior. Conversely, it may be possible to determine if an apparently aleatory series of outcomes could be produced by a system of equations for strange attractors, and also to determine the minimal number of variables those equations would link—without being able to identify the variables or verify the hypothetical identification of the system.[4]

Identification loses its relevance here, since it does not enable the prediction of observations, and observations do not enable it to be constructed. The question then arises as to what is meant by "understanding" a system of this kind. For example, there is no doubt that understanding a meteorological phenomenon "involves" the gas laws, because atmospheric variations are linked to changes in pressure and temperature. But that "involvement" becomes problematic to the extent that the notion of strange attractors is relevant to meteorological phenomena. This is because the regularity of the behavior of gases in experimental surroundings does not guarantee the regularity of the behavior they help determine in the atmosphere. The decomposition of the phenomenon into simpler phenomena does not necessarily generate *relevant* questions. The identity of the "meteorological object" no longer flows from other more "fundamental" sciences, but is defined as *singular*.

This is a first example of what can be called a complex situation: the cartography of knowledges and problems loses its "tree" shape, rising from a relatively simple but fundamental "law" or ("laws") to its application in increasingly complicated situations. The tree is a hierarchical representation: passing from the fundamental "trunk" to the tiniest branch should ideally pose technically complicated questions, but not fundamental ones. In practice, it goes without saying that knowledge of the "branches" includes a conception of the trunk, or at least of the way to pass from the trunk to the branches (for example, the quantum mechanical "explanation" of the properties of chemical bodies would have been entirely impossible without prior knowledge of their classification according to Mendeleyev's table). The finished operation, however, leaves few traces of this. To understand is to understand how the trunk generates the branches, and that is what is learned and transmitted by specialists.

Thus "strange attractors" are not a model, but rather a question mark, or an alarm bell. They signal that the difficulty of an operation of passage may not be due to a lack of knowledge, an incomplete formulation of the problem, or the enormous complication of the phenomenon, but may reside instead in intrinsic reasons that no foreseeable progress could gainsay. They signal that in certain cases more powerful computers and more numerous and precise measures could well prove useless. At the same time, they oblige us to think of the map of problems as an account of local explorations, of discoveries of possibilities of passage that prove nothing beyond themselves, that authorize neither generalization nor method.

Let's take a second example, this time concerning the very definition of a system according to its regime of activity.

Physicochemical systems involve billions and billions of molecules in interaction. A macroscopic state defined by a stable pressure or temperature is, in fact, the result of a gigantic number of molecular events that average each other out, small fluctuations aside. The relations between pressure, temperature, chemical composition, and so on represent not only what we can know about the system, but also everything that is relevant to it at equilibrium. They allow us to foresee how a state will be transformed if the value of one of its parameters is modified. Now, the judgment that distinguishes between what is relevant and what is insignificant cannot be prolonged without precautions: it depends on the stability or instability of the macroscopic state in relation to the fluctuations.

Far from equilibrium, fluctuations may cease to be noise, instead becoming actors that play a role in changing the macroscopic regime of a system. Furthermore, the far-from-equilibrium physicochemical systems that Ilya Prigogine baptized "dissipative structures" exhibit another new property. It is not only "molecular noise," the fluctuations, that may "take on meaning," but also certain details of the "control variables" that correspond to the experimental definition of the system under study (pressure, volume, temperature, flow of reagents...). For example, although gravitation has no observable effect on chemical systems at equilibrium or near to equilibrium, far from equilibrium its effect can be amplified so that it has macroscopic consequences. The system has become *sensitive* to gravitation. Similarly, it has been shown that a dissipative structure fed by chemical flows that are not perfectly constant in time but slightly irregular has access to new types of structuration. In other words, it is *the collective regime of activity* that decides what is insignificant noise and what must be taken into account. We do not know a priori what a chemical population can do, and we can no longer tell once and for all the difference between what we must take into account and what we can ignore.[5] Notice that this is also the lesson of the mathematics of catastrophe theory: only if one knows all the catastrophes to which a system is susceptible can one define the mathematical being that it represents. By trusting in appearances and forgetting the risk

this entails for the method, we can be completely mistaken even in the very definition of the system we are dealing with.

Complicated and Complex

Let's return to the contrast between complexity and complication. The references
to gods and demons that populate physics texts indicate a judgment in terms of *complication*. Laplace's demon allows one to *judge* phenomena that apparently require a
probability treatment. If we were this demon, we could understand a complicated
and apparently aleatory phenomenon in terms of deterministic laws, as in the case
of the solar system. Similarly, Maxwell's demon allows judgments to be made about
irreversible phenomena. If we were capable of manipulating individual molecules,
we could "go beyond" irreversibility, imposing an evolution that moves a system
away from its final attractor and re-creates "irreversibly" leveled differences. In this
scenario, probability and irreversibility refer to the *complicated* real with which we
are dealing, and to the approximations that limit us—but which can nevertheless
be identified by reference to the capacities of our demonic alter ego. For example,
it is said that irreversibility does not belong to the "*objective*" truth of a phenomenon, but is only relative to us. Defined in this way, the reference to "complication"
implies a more or less implicit dualism that entails the rejection and enclosure within
the domain of "nonscientific" or "simply subjective" of anything that cannot be reduced to the canon of the "simple" model.

The question of complexity arises when the relation of similar-
ity represented by the operation of prolongation between us and this alter ego starts
to pose problems. This happens in the case of "strange attractors," due to their sensitivity to initial conditions: the slightest perturbation has inordinate consequences.
A demon that understood and could control with positively infinite precision a
system characterized by such an attractor could obviously deal with it as just another system. For the demon, the system would be deterministic, as are the equations that describe it. However, is this reference still relevant? We are not actually
separated from the demon by a quantitative lack (we observe and manipulate *less
well*), but by a qualitative difference: as long as our observations and manipulations
do not have a *strictly* infinite precision, we are dealing with a system with nondeter-
ministic behavior.

The situation is similar for the unstable systems now studied by
mechanics. The deterministic and reversible trajectory that we can calculate for *simple* systems (two bodies in interaction) would require, for unstable systems, a mode
of knowledge that would only make sense for the One who knew the positions and

speeds of the entities in interaction with an infinite precision (an infinite number of decimals).[6] That being the case, is it relevant to extend to unstable dynamic systems the ideal of knowledge represented by a deterministic and reversible trajectory? Should we judge as a simple approximation the probability treatment that we *have to* apply to unstable dynamic systems, that is, judge it in the name of a knowledge that for intrinsic and noncontingent reasons we will never have?

In *Entre le cristal et la fumée*, Henri Atlan also links complexity with lack of information.[7] For Atlan, a complicated system is a system whose structure and principles of functioning are understood, and in principle, with enough time and money, nothing would prevent us from having complete knowledge. On the other hand, a complex system would be one of which we had a global perception, in terms of which we could name and describe it, all the while knowing that it was not understood in detail. For example, to the extent that a living individual is immediately perceived as organized, we have to speak about it in terms of complexity, and at the same time in terms of *lack of information*, whereas, faced with a "heap of molecules coming from a decomposing corpse," we will see no complexity unless, "for some reason, we wanted to reproduce this disorganized heap."

Here the whole issue is whether or not Atlan's global perception, which introduces the real situation of the observer, is taken as arbitrary, *whether the idea of a being interested in a living body* rather than a heap of molecules proceeds from a unilateral decision that has no correlate in the notion of complexity. If we are interested in the corpse, will we find the same tension between what interests us and what we ignore? Is the notion of complexity purely negative, in the sense that it teaches us nothing, since it presupposes a human interest *imposed* on a reality that is totally independent of this human interest?

The lack of information with which Atlan tries to define complexity can only avoid a purely subjectivist reading if it refers to our *relational* situation: the information that we possess about physicochemical interactions and that, by extension, allows us to define the "disorganized heap" as "only complicated" has meaning only in *relation* to questions about systems that are thought to require a purely physicochemical mode of understanding. There is true "lack" in the case of the living because this information, and thus this kind of relation, have, at least partially, lost their relevance: they do not allow us to give meaning to the questions that make explicit our global perception of the "living body."

Let's return to the question of unstable dynamic systems and the probabilistic description to which they limit us, we who are not Laplace's demon. This probabilistic description allows us to make sense of an irreversible behavior that is observable, that is, which corresponds to experimentally relevant questions. For the notion of complexity to have a positive meaning presupposes that these new questions are from now on the "right questions." Consequently, the simple model, which allows a general judgment in terms of reversible deterministic behavior, would here lose its status of general model, representative of dynamic systems in general (and referred back to the One for whom dynamic systems are all alike), to become a singular model suitable only for stable systems for which the difference between finite information and strictly infinite information is without qualitative importance.[8] Likewise, the existence of strange attractors makes the class of normal attractors and the judgments that they authorize appear retrospectively as singular. Here again, the possibility of once and for all identifying the relevant control variables for a physicochemical system expresses retrospectively the singularity of equilibrium and near-to-equilibrium situations.

Here the notion of complexity is close to that of *emergence*. Dangerously close, moreover, if, as is often the case, "*emergence*" is understood as the appearance of the unanalyzable totality of a new entity that renders irrelevant the intelligibility of that which produced it. In both cases, don't the "objective" categories tied to the simple model give way to *qualitatively new* questions, to categories of understanding that imply intrinsic properties having no equivalent in the simple model? We are so close to the notion of emergence — and so far. In both cases, one "starts" from a simple situation, and one describes a qualitative transformation that corresponds to the now problematic character of an operation of extension or prolongation. But the notion of emergence implies a *physical* genesis of the new, whereas the notion of complexity would correspond to a *conceptual* genesis: conceptually, we are rooted in a tradition that has given us access to a simple model and defined the tools that are appropriate for the corresponding systems. The qualitatively new questions that eventually become possible do not emerge through complexification; they express the limited character of the conceptual tools that were appropriate for singularly simple cases but that cannot be prolonged with relevance.

Notions such as emergence or complexity are nothing apart from the intentions of those who use them. The notion of emergence has too often set down prohibitions: for example, that the whole is not equivalent to the sum of its parts implies that the study of parts can teach us nothing about the whole, and furthermore that the specialists of this whole have the right and liberty to ignore any method or approach to their object that does not respect its character of self-signifying totality. As for the notion of complexity, it sets out *problems* — we don't know a priori what "*sum* of parts" means—and this problem implies that we cannot treat, under the pretext that they have the same "*parts*," all the "*sums*" according to the same general model.

The Complexity of the Living

I have thus far been emphasizing the "discovery of complexity" and thereby highlighting an example where the scientific approach has played the principal role: it is scientists who ask the questions, and complexity arises when they have to accept that the categories of understanding that guided their explorations are in question, when the manner in which they pose their questions has itself become problematic. But the question of complexity also leads to that singular category of objects that must be called *historical*, whether we are dealing with the living or their societies. In this case, the difficulty raised by the usual scientific approach, which goes from the simple to the complicated, is well known: general factors can help one retrospectively shed light on the history of a town, but one cannot deduce that history from those factors. "Complex objects," dissipative structures, catastrophic objects, and strange attractors have raised the hope of "better modelizations" for such cases. But any local success that may come of such models would only signify the identification of *simple* aspects of these historical objects. Here again, complexity is not a theory, an exportable general model. The "complex" lesson of dissipative structures is not the appearance of coherent collective behaviors but this gravitational factor that, according to the circumstances, can be insignificant or "change everything."

One cannot speak of a "discovery of complexity" with respect to historical objects. Indeed, the notion of complexity is usually recognized as almost constitutive of the "living object." The question, then, is to understand how and at what price the *problem* of complexity is or can be integrated with the approach of the biologist.

It is not a matter of denying that certain experimental questions concerning the living are "simple"—this is the case, for example, when it is a question of clarifying a physiological relation whose logic plays an essential role in the stability of a living behavior. In this case, significantly, the experimental question expresses an intrinsic finality of the relation in question. Given the function of the muscle, how is this function realized? Given the function of transmission of the neuron, how and in what circumstances does it transmit? And so forth. Here the questions of experimenters are stabilized by the very *role* that they attribute to their objects in the organism. From the moment biologists have understood which function resolves which problem, they can ask themselves *how* the problem is in fact resolved. But, even when these "simple" questions are dealt with, they give rise to another type of problem: the question posed by the living is not simply the question of knowing *how* it realizes the different functions necessary to the survival of the organism, but also the question of knowing how to understand, at the phylogenetic level as much as at the ontogenetic level, the production of these functions, the production of *meaning* that experimenters subscribe to when they ask "how."

At this double level, both phylogenetic and ontogenetic, we encounter a dominant model, based on the pair "simplicity/complication." The history of the living (mutations, selection of the best "adapted"), as well as the development of the living individual, would follow *simple* principles that would nevertheless generate appallingly complicated histories. This is what is implied by Jacques Monod's remark that what is true for the bacteria is true for the elephant, and its implicit correlate, for which his *Chance and Necessity* gives the theoretical justification:[9] the bacteria is the simple model that gives meaning to the set of instruments that will enable us to understand the elephant (or man).

Here again, the problem is that of representativeness. Is the object "bacteria" a singular borderline case, or rather, is it representative of the living? In particular, does the fact that bacteria do not develop—with a double consequence that constitutes them as privileged experimental objects: their study in vitro presents no problems, since the only question is to know whether or not the surroundings contain the nutritive products necessary for their multiplication; the relations between genetic instructions and metabolic performance are here open to direct genetic-biochemical exploration—produce an obstacle to their definition as the royal road of intelligibility for organisms that have an embryonic development?

Chance and Necessity is an important book in that it links in an explicit way the thesis of molecular biology concerning the living to this problem of representativeness. Both the conception of ontogenesis and that of phylogenesis are actually based on a crucial affirmation: *like the bacteria*, every living being can be assimilated to a process of *revelation* of the genetic program; that process is certainly abominably complicated, but its principle is clear. As a result, explanation in biology splits into two essentially autonomous approaches: the complicated description of physicochemical processes that actualize the informational content of the genome and render possible the construction and functioning of the living; and the reference to a complicated history during the course of which natural selection created and modeled this content.

This distribution of the explanation has the particularity of con-centrating the singularity of the living on a single mechanism: that of natural selec-tion. Here, natural selection is uniquely responsible for the fact that biochemical processes result in the constitution of an organized being that, apparently, is gov-erned by a finality: to survive and reproduce. Reciprocally, teleonomy, the apparent finality of the living, is clearly only an appearance, but it expresses, at the level of the living individual, the only raison d'être for what biochemistry, physiology, or embryology describe. In this hypothesis, the living would be integrally formed by the selective constraints of material arbitrarily produced by the mutations of genomes. Thus "teleonomy," the fact that the living *appears* as if "made in order to repro-duce," finds itself in the same dominant position from the point of view of the ex-planation as the final causes invoked by Aristotelian biology: selection gives to the living its only conceivable meaning.

However, if bacteria are not the representative model, natural selection loses its status of "cause" in the final analysis. Selection plays on differ-ences among organisms, and expresses itself in a change in the genetic composition of the population. If the relation between genetic information and organism is not the "revelation" of which bacteria provide the operational model, the question arises of knowing "how" selection can evolve structures that are genetically constrained but not genetically determined.

This is what the British biologist Conrad Waddington clearly saw when he introduced the notion of "canalization."[10] Waddington starts with the idea that, as a general rule, the development of the living organism should not be thought of as a revelation but as a construction that integrates genetic constraints and interactions with the surroundings. Selective pressure can, by an accumulation of genetic constraints, progressively canalize the path of development of certain traits. Given this, development, *insofar as it concerns these paths*, will indeed appear as a "revelation" of the "normal" consequences of the genetic information.

Here again, questioning the simple model implies a certain re-versal of perspective. Usually, one derives a kind of fundamental notion of the liv-ing from the stereotyped, informationally closed organism. It follows that, as they become more complex, certain living forms acquire, come what may, certain possi-bilities, apparent or actual, of learning and open behavior. In Waddington's perspec-tive, the spectrum becomes horizontal rather than hierarchical. Living organisms, such as bacteria, which can be considered on a first approximation as closed on them-selves (from an informational point of view), lie at one extreme of this horizontal spectrum, and, on earth, human beings lie at the other. Admittedly, the spectrum can only be actualized progressively during the course of evolutionary history, but the problem for which it constitutes a set of solutions is posed at the same time as the problem of life. The stereotyping, as well as the openness, has to be explained.

Selective pressure stabilizing the development of certain characteristics and accumulating constraints that favor their actualization submits such a development to a stereotyped norm and thereby creates a *temporality of a repetitive type*: the living organism repeats the "species." The risk is taken, in this selective invention of paths of development that can be foreseen and reproduced, of depriving the ontogenetic process of its sensitivity to circumstances, and thus of the possibility of an innovative and eventually interesting response to a modification of the surroundings. Therefore, it is necessary to understand that the degree of canalization or, on the contrary, of openness to circumstances and variations of the surroundings constitutes, for each phenotypic trait, a *wager* with regard to the ecological circumstances that will affect the future of the population.

This wager—stereotyped or open?—entails a consideration of what an individual is within a population, that is, of what ecologists call the strategy of a population. For example, one might say that certain species of parasite have made the wager of stereotyping: a parasite lives its life cycle as is, and in so doing, it repeats, if it succeeds in reproducing itself, a *specific* behavior. There is not much sense in thinking that a parasite learns or adapts. One might say that, from the point of view of the strategy of the population, the individual in its case is both "all" and "nothing." The notion of a parasite as an individual is coextensive with that of a parasite as a species; within each generation, only a tiny proportion of parasites will survive, but their rate of reproduction is sufficient to ensure the survival, indeed the proliferation, of the population. Selection, which in this case is ferocious, appears as if free and all-powerful, endlessly finicking over the superb reproductive automaton constituted by each individual. On the other hand, a bird, a chimpanzee, or a human being *learns*. The behavior of the individual does not repeat the species since each one constitutes a singular construction that integrates genetic constraints and the circumstances of a life. Furthermore, selective pressure does not bear on the individual but on the individual in its *group*, in the strong sense: it is not a question of knowing how an individual will "take advantage" of its group (the thesis of sociobiology). The group has become the condition of possibility for the individual, whose development involves protection, learning, and relations.[11] The individual now appears as a sheaf of linked temporalities. It cannot be understood simply as a function of the "species memory," constituted by its genetic constraints.

It must also be understood as the memory of its own experiences, indeed, ultimately, for human beings, as an indefinitely multiple memory of all the pasts we have inherited and to which we are sensitive. Here, genetic constraints, like the notion of species, take on a quite abstract meaning in comparison with the notion of the concrete individual.

The Risk of Complexity

Here again we come up against the problem of the relevance of the perspective. The eventual limits of the relevance of descriptions made in terms of genetic determinations, or, more generally, made using laboratory methods of isolation, are not in the first instance tied to ideological or humanist choices, nor to the problem of the complication of the living object. They are intrinsic and irreducible because they involve the distinctive temporality of the object studied. The privilege of the bacterium, the possibility of isolating and studying it in vitro, expresses the singular, determining role, played in its case by genetic constraints. In other cases, isolation is a dangerous game, and those who believe they can purify their objects in fact intervene actively in the significance of the object they observe. In a general manner, one can say that the quasi-paranoiac precautions that (for example, in experimental psychology) ensure the reproducible character of experimentation, emphasize what, through a methodological concern for purification, these observations wish to neglect: precisely the fact that the behaviors of the beings under study are not purifiable from their context. In the final analysis, from the moment that the experimentation is not addressed to an established fact but to a *being produced by history and capable of history*, it is addressed to something that is certainly not a subject in the human sense of the term, but is not a pure object either. No "methodology" can decide, in the name of the "constraints of scientificity," to deny, through the way in which it defines its interrogation, the fact that, to varying degrees, the sense of its interrogation may also present problems for that which is interrogated.

The intrinsic complexity of living systems—the fact that they are the product of multiple histories in relation to which all constraints (genetic, experimental, or otherwise) take on meaning—does not impose a dramatic limit on any possibility of experimentation. What it imposes is the necessity for an intelligent experimentation, which assumes the risky responsibility of asking relevant questions. Every question is a wager concerning what the interrogated object is sensitive to, and no method is neutral with respect to this problem. The problem of relevance does not lead to irrationalism, but to the ever-present risk of "silencing" the very thing one is interrogating.

In this situation, to speak of the "discovery of complexity" may appear paradoxical. Indeed, the problem I have just described is not a new one. It was underscored by eighteenth-century thinkers such as Diderot and Lichtenberg. If there is an "event" here, it does not relate to a general history of knowledges, but to the concrete history of the sciences and the notion of scientific discipline as invented by the academic institutions of the nineteenth century.

I have spoken of the risk tied to the double separation, conceptual and technical, that makes experimentation possible. But the nineteenth century invented and put into place a third separation, to which present-day science owes its characteristics: the social separation between those who "know how to recognize the facts" and those who are incompetent and only have opinions. To the discipline — understood not in the simple sense of specialized research but in the sense that this specialization implies a position of authority over the definition of "scientific fact" — there corresponds the pair "simple/complicated." This is the notion of an approach that guarantees its own scientificity and defines the knowledges to which its models seem unable to give meaning, as "opinion," as fallow land waiting for the prolongation that would finally endow it with meaning.

In this respect I refer to Judith Schlanger's beautiful book *Penser la bouche pleine,*[12] which poses the problem of the fascination exerted on those who analyze and organize by the products of their own distinctions. She observes that this fascination is not total. Egyptologists distinguish their object, "Egyptologizable" Egypt, but in the very language they employ, *nourishing their interest for this Egypt,* there coexist many other Egypts, which they know have also fascinated: the Egypt of the Greeks, the Egypt of myths, the Egypt of novels and films. Schlanger argues that it is this "cultural memory," the knowledge that other self-evidences concerning our object have existed and still exist, that *reintroduces the world between us and ourselves,* preventing us from fully adhering to theoretical self-evidence. It is this cultural memory, conveyed by words or by the effective coexistence of knowledges, that maintains the remembrance of risk, giving a meaning and measure to relevance, and favoring, in certain cases, theoretical innovation. However, the notion of discipline as it is used by modern research institutions coincides with the invention of a form of education for scientists that renders this "cultural memory" as empty and trivial as possible. This is demonstrated by Thomas Kuhn when he describes how disciplinary paradigms are inculcated into apprentice specialists.[13]

Quite obviously, if the pair "simple/complicated" has been stabilized by modern research institutions, complexity cannot in and of itself be synonymous of a reunion with an open practice of science. No "discovery of complexity" per se is capable of challenging academic isolation and the experimental mono-logue that it stabilizes. The response to the question of complexity is not theoreti-cal but practical. It requires what Jean-Marc Lévy-Leblond called "the enculturation of science." Let us retain the boldness that gives the experimental question its beauty and interest, while eschewing the recklessness so often claimed today as a condition of this boldness. Let us take, accept, and learn to measure the risks.

T W O

Breaking the Circle of Sufficient Reason

FOR HISTORIANS of science, the "breaking of the circle" refers to a precise episode: the audacity of Kepler, who was the first who dared to free himself from a conviction that had guided astronomers, from Greek antiquity up to Copernicus, and even to Galileo. All these astronomers had accepted as self-evident that to the "perfect" (regular and eternal) movement of the planets there should correspond the perfect geometrical figure, the circle. We are used to associating the birth of modern science with the "Copernican revolution," the substitution of the heliocentric system for the geocentric system. But it was Kepler who made the real difference between the two systems by transforming the significance of the relation between mathematics and astronomy. The circle, the mathematical figure of perfection, enabled one to judge a priori the observable world: the technical problem for astronomy was to "save the phenomena" by simulating in the language of circles the movement of celestial bodies. Kepler divested mathematics of this power of judgment and used mathematics as a research tool, thereby arriving at a figure that was for him just one among others: the ellipse. As with Freud, one could speak here of a wound imposed on human narcissism. Reason had to relinquish the power of judging a priori and submit to empirical observation.

Of course, with Newton ellipses ceased being just one geometric figure among others; they were given their theoretical meaning as the expression of the gravitational force between the sun and the planets. The wound opened by Kepler only deepened, causing a scandal among rationalists such as Huyghens, Leibniz, Euler, and d'Alembert: what is this force of attraction acting instantaneously at a distance? More than two centuries later, with Einstein's theory of general relativity, the scandal was finally quelled and the force of attraction now expresses the metric properties that Einstein attributed to space-time. But other forces had in the meantime made their appearance. We know the passion with which today's physicists, such as Stephen Hawking, continue the quest undertaken by Einstein, the unification of physical forces. Isn't this still the same passion? Isn't it the same desire to rediscover a rational world that is intelligible a priori, to close the wound opened by Kepler's ellipses? Isn't it the desire to reconstitute the broken unity between the powers of reason and the reasons of the world?

This initial history can be viewed as a parabola that adequately demonstrates the singularity of what we call physics. On the one hand, this is clearly the science where the relation between theory and experience is the most rigorous and demanding, and in that, physicists are clearly the descendants of Kepler. But, on the other hand, this is a science that always appears to involve the project of *judging* phenomena, of submitting them to a rational ideal. More precisely, we are dealing with the only science that makes the distinction between what physicists call "phenomenological laws" and "fundamental laws." The first may well describe phenomena mathematically in a rigorous and relevant way, but only the second can claim to unify the diversity of phenomena, to go "beyond appearances." It is as if, instead of being abandoned with Kepler, the idea that we have the right to judge the world of phenomena in the name of a normative ideal had become more virulent than ever.

It was by working with Ilya Prigogine that I learned to become sensitive to this singularity of physics, to understand it as a problem. When I was a student, it was transmitted to me as self-evident. My understanding was able to develop because the problem that occupies Prigogine is, and always has been, the problem of time in physics. It is with respect to time that the distinction between phenomenological and fundamental takes its most dramatic and polemical turn. It is in relation to time that the rational ideal that appears to guide physics has the most paradoxical consequences.

When I learned physics, I accepted as "only phenomenological" the laws that describe "irreversible" evolutions—that is, evolutions that only take place in one direction (a mixture does not "unmix" itself; temperature differences do not increase spontaneously; a vacuum is not spontaneously produced in some corner of the atmosphere). From a fundamental point of view, the difference between the evolutions that we observe and those that we think impossible is not valid. If I describe the leveling of temperatures in terms of the movement of molecules and their collisions, I have to conclude that for each particular evolution of molecules that would be expressed at the observable level by a leveling there corresponds another, perfectly equivalent, which, in its case, is expressed by an increase in temperature difference! The fundamental laws of physics do not recognize what leads us to recognize without hesitation that from one situation another will follow. It gives no direction to what we traditionally call the "arrow of time."

When you are studying physics, you quickly get used to accepting things. But none is as mind-boggling as this. Never has any speculation, however audacious, whether of mystics or philosophers, attained such violence: to judge as "only phenomenological" the difference between before and after! If one showed astronomers a film representing an unknown planetary system, they would be unable to say whether the film was projected from beginning to end or from end to beginning. And it is this same incapacity that physics attributes to the "perfect observer," for whom nature would be directly decipherable in terms of fundamental laws!

Yet the sun shines, and without it living things would not exist, and if the measuring instruments of physicists, or even only their retinas, were not irreversibly marked, there would not be any measuring, and physics would be impossible. Is it sufficient to denounce the absurdity of a physics that states that its own *conditions of possibility* are relative to the approximate description that can be made by the imperfect observers that we are? Can we quite simply denounce the ideological character of this type of judgment and turn the page?

Something I also learned with Prigogine is that the page is not so easily turned. It is not that simple to make the distinction between serious physics and its ideological dimension. Physicists do not just affirm an ideological judgment; their work has been actively and creatively guided by what we are tempted to call "simply ideological" judgments. One could even say that the theoretical tools they used conveyed this judgment *even before they became aware of it*. The most astonishing singularity of the question of time in physics is that it is only at the end of the nineteenth century, through Ludwig Boltzmann's work, that physicists realized that the laws they had taken for granted for about two centuries and accepted as fundamental did not allow them to distinguish before from after! They knew from the beginning that these laws were deterministic, that they treated uncertainty regarding the future as an effect of our ignorance. They had not seen until then that these laws entailed a radical difference between two types of deterministic laws—for example, on the one hand the law that describes a perfect pendulum, oscillating eternally without friction, and on the other hand Fourier's law, which describes how heat diffuses between two points of different temperature. It was only when Boltzmann tried explicitly to join these two types of deterministic laws and came up against paradoxes judged as insurmountable that physicists understood the choice that was forced on them: either to make the arrow of time an appearance or clearly admit that the laws of movement did not constitute a privileged access to the labyrinth of phenomena.

We have tended to forget the late-nineteenth-century controversies that marked this episode. They are reminiscent of the controversies that have surrounded quantum mechanics for more than half a century now. In both cases, it was the very meaning of physics, its vocation and the ideal that guides it, that were under discussion. We have forgotten these controversies because the revolutions of twentieth-century physics—relativity and quantum mechanics, which are the direct descendants of dynamics—appeared to confirm the choice of those who sacrificed the phenomenological evidence of the arrow of time to the laws of dynamic movement. Be that as it may, at the time physicists chose Einstein over Kepler. They chose a physics that judges in the name of a norm, as opposed to a physics that accepts the wounds that phenomena imposed on its ideal. This is doubtless the source of the renewed power of the dream I alluded to earlier: that of a physics that, thanks to the unification of forces, would finally "heal" the Keplerian wound, rediscover the path of a rational cosmos, and find a point of view from which phenomenological diversity could be judged rather than endured.

One might also add that the paradoxical consequences of this choice are still present in the paradoxes of quantum mechanics, whose formalism in fact links the two contradictory times of physics: the symmetrical time of dynamics and the irreversible time that quantum mechanics associates with the operation of measurement. It is because physicists have not managed to submit the second to the first that some have taken the position that quantum mechanics involves an intervention of human consciousness. A measurement, they say, is not a physical process that the physicist uses to observe and know; it is human awareness of a measurement that actively creates the observable phenomenon.

Breaking the circle of sufficient reason: the title of this essay refers both to the tradition of physics and to a possible future, one that, it seems to me, Prigogine is working toward. This is a future that would replay the choice that led physicists to define physics against the phenomenological evidence of the arrow of time. Prigogine sees himself as a true descendant of Boltzmann, who tried to join the laws of dynamic movement and irreversible processes. For me, Prigogine is also clearly a descendant of Kepler, who dared to break with the perfection of the circle and to wager on the *relevance* of mathematics to describe the world of phenomena, against the *power* of mathematics to judge this world in the name of a normative ideal.

The principle of sufficient reason is not physical but philosophical. It derives from Leibniz, who baptised as "dynamics" the science of movement created by Galileo and continued by Huyghens. Leibniz understood that this science owed its fecundity to the discovery of a new mode of access to phenomena, to a new manner of understanding and describing them. Leibniz stated that "the full cause is equivalent to the entire effect." In modern physics, the perfection of a science that can define this quantitative equivalence between cause and effect has been substituted for the perfection of the circle. I will show that it is this perfection of science that defines the new normative ideal on the basis of which physics has come to deny the difference between the past and the future.

Let's imagine Galileo confronting his problem: movements that need to be compared and brought together under one law, even though they involve bodies that move at continually changing speeds, in different time frames, over different distances. I can compare the speeds of two bodies if they cross the same amount of space, or if they move for the same amount of time; or, I can compare the distances and times if they travel at the same speed. But if all three terms are variables, what do you do? We know that Galileo worked on this problem for years without getting anywhere. The solution he eventually found, concerning bodies falling vertically or otherwise, is still the one taught in school, and which beginners often have the greatest difficulty in grasping. Galileo understood that if the movement is accelerating, if the speed is continually changing, he could not continue, as had all his predecessors, to define the speed as "the space traveled divided by the time taken to travel it." He invented an idea that is difficult to accept: that a body in accelerated movement changes speed at each instant. Thus, at each instant it "has" a speed, but at this speed it travels no space in no time since at the following instant it will have a different speed. Thus he defined the "velocity" at a given instant as that which the body had *gained* during the course of its fall up to that instant. But how is this gain to be defined? It is in the answer invented by Galileo that Leibniz will find the "principle of sufficient reason": the gain must be "equivalent" to a loss. The effect must be equivalent to the cause that disappears in producing it. The velocity gained is "equivalent" to the lost altitude, which, for Galileo and for Leibniz, means that *this velocity is exactly sufficient to allow the body to regain the lost altitude.* Today we write $1/2 \ mv^2 = mgh$, and it is the "equals" sign between the two that shows the equivalence between the entire effect (the velocity gained by a body of mass m, falling with a uniform gravitational acceleration g) and the full cause (the difference in altitude h covered by this fall). The proof of the equivalence is the reversibility of the relation between cause and effect: if the body rises, the effect is measured by h, and the cause is the velocity lost by the body at each instant of its rising motion.

The notions of cause and effect are old, dating from well before modern physics. What is new is the "equals" sign. It is that sign that will dominate and determine the identity of cause and effect from this point forward. Leibniz's contemporaries were scandalized. Why measure the effect or the cause by the *square* of the velocity? Why this arbitrariness? Why not make it the cube while one is at it? But Leibniz had understood that the only objective definition of a cause or an effect that is not a reflection of our choices or ideas is one that acquiesces to equality: if I define cause and effect in such a way that I can write the equals sign between them, I know that my knowledge is perfect, that I have let nothing escape.

The "equals" sign of sufficient reason is a historical novelty. The definition of the "instantaneous velocity" will henceforth join past and future under the sign of identity: the present state is defined as an effect equivalent to the past that caused it, and as the cause of an equivalent, possible future (a body, with this velocity, *could* reascend). And Leibniz was right: it is this definition of the instant that the mechanists of the eighteenth century, from Bernoulli to Lagrange, generalized. It lies at the basis of their notion of dynamic system, and it is why the equivalence between the past and the future, the impossibility of distinguishing between before and after, is written into the equations of dynamics. It is written in not as a particular result, but literally as a *syntactic rule* that defines the meaning of what we call velocity, force, and acceleration. It is why the negation of the arrow of time by physics is not only "ideological." It was not done intentionally; it was an instrument of definition.

Already at that time, the phenomenological evidence was set against Leibniz: in collisions between bodies that are not perfectly elastic, some movement is "lost," and thus the cause is not equivalent to the effect. Even in his own day, Leibniz used the argument that reigns today in physics: it is because we are not perfect observers, capable of measuring the full cause and the entire effect, that we believe that some movement is lost when actually it is transmitted to small, invisible parts of bodies. It is as if we had changed a sum of money made up of notes of large denominations into loose change, and, being imperfect financiers, we were incapable of taking into account the small coins. The principle of sufficient reason articulates perfection and equivalence. It defines the manner in which an observer who lets nothing escape can observe and calculate. Only an ideal observer can write the "equals" sign. Nonequivalence — in the nineteenth century, the fact that irreversible processes do not produce effects capable of restoring what they destroy (we cannot "reverse" the process of heat diffusion and restore the difference of temperatures that determined it) — designates the imperfect observer and manipulator, incapable of having access to the fullness of the cause and the entirety of the effect.

Breaking the circle of sufficient reason? What I learned with Prigogine is that it is not enough to understand the problem in order to to resolve it. Sufficient reason inhabits the very syntax of our equations, from dynamics to relativity and quantum mechanics. Physicists cannot simply abandon these equations since it is through them that physics established a particular experimental relation with the observable world. We cannot simply abolish a history in which the laws of dynamics have played a major role. Philosophers can certainly understand how this history inseparably associates ideas, interpretations, and experimental constraints, but they cannot "remake physics." As for physicists, they are situated in a tradition from which they receive their instruments and their language: if they reject them they lose any possibility of communication with their colleagues, they are no longer physicists, they find themselves isolated, alone in a labyrinth of phenomena that have again become indecipherable.

Prigogine's choice (at least as I can retrospectively draw a lesson from it, since it concerns a complex program of research undertaken over some forty years) has been to take stock of the situation — namely, the success of physics and the role as driving force that dynamics has played in it despite the negation of time it brought about. His choice has been to pose the problem of the "approximations" physicists use to rationalize that the "fundamental laws," which ignore the arrow of time, actually explain the "phenomenological descriptions" — all of which affirm the arrow of time. He has tried to show that these "approximations" in fact express intrinsic properties of the objects described by phenomenological laws. In other words, he has tried to give a positive, objective meaning to that which, since Leibniz, physicists have attributed to the imperfection of the observer. Can we define an "intrinsic difference" between, on the one hand, the movement of the moon, the movement of a perfect pendulum, and the trajectory of a falling body — for which sufficient reason is relevant — and, on the other hand, a population of 10^{23} colliding particles — which forces a recognition of the phenomenological evidence of the arrow of time? By "intrinsic difference" is meant a difference that cannot be reduced to the incapacity of the observer to define the movement of each of the 10^{23} colliding particles.

By raising such questions, Prigogine has thus sought to challenge the tradition from the inside, to transform the link between the fundamental and the phenomenological.

Sometimes there are happy coincidences in the history of physics. One of them is certainly the encounter, almost two decades ago, between Prigogine's pursuit of this problem in relative solitude and a spectacular development in dynamics—namely, the study of unstable dynamic systems, and in particular of *chaotic* systems. What is a chaotic system? It is a highly unstable system, one so unstable that, starting from quasi-identical initial states (described by the same numbers, but only up to the tenth, or the hundredth, or the thousandth decimal), it will deploy trajectories that, instead of remaining close, very rapidly diverge from each other. At the end of a very short period of evolution, systems that were originally indistinguishable will thus have entirely dissimilar behaviors. The movements of the moon or a pendulum are not chaotic, but the movement of 10^{23} particles is a chaotic system.

Why would chaotic systems enable us to break the circle of sufficient reason? After all, they are defined by the equations of dynamics, like all other dynamic systems, and thus in principle by the equality of the full cause and the entire effect. But is this principle still legitimate? It is from that angle that Prigogine attacked the problem. When a system is chaotic, only a God, capable of defining the instantaneous state of a system with infinite precision would be able to calculate the trajectory, deploying the succession of causes and equivalent effects. But the physicist is not God. He or she must observe and measure, and every measurement, every observation ends up with a finite number of decimals, however numerous they may be. The physicist does not have, cannot have, and will never have access to perfectly full causes and perfectly entire effects, however far technical progress is pushed. For the usual dynamic systems, which inspired the principle of sufficient reason, this is not of great importance. If the approximation is good, it stays good. The observer who can define a "nearly full" cause and a "nearly entire" effect defines the trajectory "nearly perfectly." For chaotic systems, everything changes: there is a qualitative difference between the type of description a finite observer can construct, however precise the measurements, and that of the infinitely perfect Observer. From the point of view of the One who would have the full and entire definition of the causes and effects, there would be no difference between the stable movement of a pendulum and the behavior of a chaotic system. For all finite observers, however perfect they are, this difference is irrevocable. They will have to rely on probabilities, renouncing the ability to foresee the behavior that a system will adopt, even if they know it as well as is possible.

Sufficient reason constituted an ideal of perfection because it was the ideal of a complete definition, which lets nothing escape. Thus, any divergence from this ideal could be attributed to the observer, referred to the person who describes rather than to the world described. Chaotic systems do not force physicists to abandon this ideal. Prigogine's approach has been criticized by physicists who prefer to think that a chaotic system is deterministic, even if this determinism must appeal directly to divine knowledge. But chaotic systems mark the point where the ideal of sufficient reason *can* be abandoned without arbitrariness, where this abandonment does not signify a renunciation of a "better knowledge that is in principle possible." It is on this point that the physicist can invent a new dynamic language, which brings to light the intrinsic character of the phenomenological difference between stable and chaotic dynamic systems. As Prigogine has shown, the price to be paid by a relevant dynamic description of chaotic systems is precisely that this description gives an objective, intrinsic direction to the arrow of time that characterizes phenomenological descriptions. The description of chaotic dynamic systems, freed from an ideal of sufficient reason that is not pertinent to them, breaks the symmetry between before and after imposed by that ideal.

Breaking the circle of sufficient reason, giving meaning to phenomenological distinctions, posing the question of the relevance of our concepts — all this relates not to a description of the world in itself, as dreamed of by the physics of sufficient reason, but to human undertaking, as creative of meaning. Ultimately, this is what Prigogine really taught me: to be able to respect the activity of physicists without having to believe that it was neutral, that is to say, divested of passion, subjected to a reality that would be capable of dictating the manner in which it must be unraveled. The negation of time by the physics of sufficient reason does not express a simple "ideology" of which physics could easily be purified. It also expresses a passion, that of escaping the arbitrary. This passion has been the source of invention, even if it ended up defining as illusory a distinction without which the practical activity of humankind loses its meaning. Prigogine's passion has been to attempt to give a new meaning to the refusal of the arbitrary that guides physics, to identify physical systems that would allow one to give an objective sense, valid for all conceivable observers, to what physicists had until then defined as "approximation."

I wrote this text with Kepler as its dominant theme because it was he who transformed the sense of mathematical astronomy by "breaking" the circles in the name of which the phenomenological evidence of the celestial movements was judged. The Keplerian adventure is exemplary but not unique. The passionate maintenance of points of view that satisfy our ideals, as well as the passion of "breaking the circle," of recognizing and questioning the theories that put the power of judging before the requirement of relevance, inhabit not only physics but all areas of our knowledge. This is why the sciences must not only be seen as expressing reason as authority and judge, but also be understood as expressing reason as adventure.

T H R E E

The Reenchantment of the World

(with Ilya Prigogine)

The End of Omniscience

SCIENCE IS certainly an art of manipulating nature. But it is also an attempt to understand it and respond to questions that have been asked by humankind generation after generation. One of these questions has become like an obsessive theme throughout the history of the sciences and philosophy. It is the question of the relation between being and becoming, between permanence and change.

Pre-Socratic speculations were marked by a number of decisive conceptual choices: Is change, whereby things are born and die, imposed from the outside on matter that remains indifferent to it? Or is change the product of the intrinsic and autonomous activity of matter? Is it necessary to evoke an external driving force, or is becoming immanent in things? During the seventeenth century, the science of movement arose *in opposition to* the biological model of a spontaneous and autonomous organization of natural beings. From then on it found itself torn between two fundamental possibilities; for if all change is nothing but movement, what is responsible for this movement? Is one obliged, like the atomists, to rely on atoms in the void along with their chance collisions and precarious associations? Or is there a "force" exterior to masses that is responsible for their movement? In fact, this alternative raised the question of the possibility of giving a lawful order to nature. Is nature intrinsically aleatory, and are regular, foreseeable, and reproducible behaviors simply the product of fortuitous chance? Or, rather, does law come first? Can we define the forces imposing on inert matter a lawful behavior, susceptible to mathematical description, as the very principles of physics?

In the eighteenth century, the uncertainty of precarious and spontaneous whirlwinds was conquered by unchanging mathematical law; and the world governed by this law was no longer the atomistic world where things were born, lived, and died in the risks of an endless proliferation; it was a world in order, a world in which nothing can be created that had not always been deducible from the instantaneous definition of any of its instantaneous states.

In fact, the dynamic conception of the world does not constitute in itself an absolute novelty. Quite to the contrary, we can situate very precisely its place of origin: it was the Aristotelian celestial world, the unchanging and divine world of astronomical trajectories, which was, according to Aristotle the only world that could be given an exact mathematical description. We have echoed the complaint that science, and physics in particular, has "disenchanted" the world. But it disenchants it precisely because it deifies it, because it denies the diversity and natural becomings that Aristotle attributed to the sublunar world, in the name of an incorruptible eternity that alone could be truly thought. The world of dynamics is a "divine" world untouched by time and from which the birth and death of things are forever excluded.

However, this apparently was not the intention of those we call the founders of modern science; in challenging Aristotle's claim that mathematics ends where nature begins, it seems that they did not intend to discover the unchanging behind the changing, but rather to extend a changing and corruptible nature to the boundaries of the universe. At the very beginning of his *Dialogue on the Two Chief World Systems*, Galileo is amazed that some people could think that the earth would be nobler and more admirable if the Great Flood had left only a sea of ice behind or if the earth had the incorruptible hardness of jasper. He concluded: let those who think the earth would be more beautiful after having been changed into a crystal ball be transformed by Medusa's gaze into diamond statues, thereby rendering them "better" than they are.

But the objects of science chosen by the first physicists who attempted to mathematize natural behaviors—the ideal pendulum with its eternal and conservative oscillation, the cannonball in a vacuum, simple machines with perpetual motion, and also the trajectories of planets, from then on assimilated to natural beings—all these objects, which were the concern of the first experimental dialogue with nature, were found to correspond to a *unique* mathematical description, a description that reproduced exactly the divine ideality of Aristotle's stars.

Like Aristotle's gods, the simple machines of dynamics are only concerned with themselves. They have nothing to learn; rather they have everything to lose from any contact with the outside. They simulate an ideal that the *dynamic system* will actualize. This system rigorously constitutes a system of the world, making no place for a reality exterior to it. At any instant, each point in the system "knows" everything it will ever need to know, that is to say, the spatial distribution of every mass in the system and their speeds. The system is always and everywhere self-identical: each state contains the truth of all the others, and each can be used to predict the others whatever their respective positions on the monotonous axis of time. In this sense, one can say that dynamic evolution is a tautology. Deaf and dumb to whatever outside world there may be, the dynamic system functions by itself and all states are alike to it.

The universal laws of the dynamics of trajectories are conservative, reversible, and deterministic. They imply that the object of dynamics can be completely understood: the definition of any state of a system, and the knowledge of the law that governs its evolution, allow one, with the certitude and precision of logical reasoning, to deduce the totality of its past and future.

Consequently, nature viewed on the model of a dynamic system could no longer be other than foreign to the one who described it. The only possibility left was to adopt the position of the optimal description, where Laplace's demon, undaunted, has always already calculated the world, past and future, after having determined, in any given moment, the values of the positions and velocities of each particle.

Many critics of modern science have emphasized the passive and submissive character that the mathematics of physics gives to the nature it describes. It is a logical consequence of this mathematics that a totally foreseeable, automatic nature would be capable of being thoroughly manipulated by those who know how to determine its states. However, we think that the diagnosis cannot be that simple. Certainly, "to know" has often been identified, during the last three centuries, with "to know how to manipulate." But that is not all there is to it, and the sciences cannot without violence be reduced to a mere project of mastery. They also involve dialogue—not, of course, exchange between subjects, but explorations and questions whose stakes are not those of the silence and submission of the other.

First of all, it is necessary to make a distinction between dynamics and other sciences where the idea of manipulation plays a role. For example,

Skinnerian psychology teaches one how to manipulate living beings, which it treats like black boxes: the only things it considers relevant are the "inputs," which it controls, and the "outputs," the reactions of the subject of experience; in the same way, the science of steam engines did not have the ambition of "entering" into the furnace, but solely of understanding the correlations between variations of magnitude measurable *from the exterior*. On the other hand, dynamics exhausts the object "in itself" with a set of *equivalences that define equally and inseparably the possibilities of manipulation*. The best example is that of the reversal of velocities. In order to identify the cause and the effect, which disappeared in determining the change, and the equivalent gain that constitutes the change, one invokes an ideal manipulation where the velocity would be instantaneously reversed. The body would regain its initial *altitude* while losing the whole of its acquired *velocity*. The fundamental equivalence $mv^2/2 = mgh$ both defines the dynamic *object* "objectively" and defines an ideally possible *manipulation*.

Thus, dynamics achieves in a singular way a convergence between the interests of manipulation and those interests of knowledge that simply wish to understand nature. It is thus understandable that science was able to seem dominated by the ambition of manipulating, but also that this domination proved to be unstable, when new objects attracted its attention and curiosity.

From this point of view, there is without doubt no better symbol of a transformation that is primarily of our questions and interests than the evocation of the two subjects that held Kant's admiration: the eternal movement of the stars, in the sky, and the moral law, in his heart; two lawful orders, unchanging and heterogeneous. We have now discovered the violence of the universe; we know that stars explode and that galaxies are born and die. We know that we can no longer even guarantee the stability of planetary movement. And it is this instability of trajectories, these bifurcations together with the bifurcations and creative risks in our lives, that are today a source of inspiration to us.

We have tried to understand the complex processes by which the transformation of our interests and the questions that we consider decisive have been able to enter into resonance with the paths of research proper to science, and also to locate within the closed coherence of its certitudes the opening that we have just recounted. And because it was a question of modifying the scope of concepts, of shifting problems into a new landscape, of introducing questions that drastically change the definition of disciplines, in short, because it was a matter of inscribing within science the urgency of new preoccupations, this opening took multiple and often sly paths.

The history of thermodynamics is perhaps, in this respect, exemplary.

We have given it as a starting point Fourier's formulation of the law of heat diffusion. This was the first intrinsically irreversible process to be given a mathematical expression, and it is this that caused a scandal: the unity of mathematical physics based on the laws of dynamics was shattered forever.

Fourier's law described a spontaneous process: heat diffuses—there is no way of preventing or reversing it, in short, of controlling it. On the contrary, in order to control heat, any conduction, any contact with bodies at different temperatures, must be *avoided*. Fourier's law describes in particular an irremediable waste while the problem at the start of thermodynamics was to employ heat to make a motor work. This is why the Carnot cycle, from which the laws of thermodynamics would be formulated, can be reduced to a set of *ruses* that seek to minimize irreversible conduction. Thermodynamics is thus set up in relation *with* irreversibility but also *against* it, seeking not to know it but to avoid it. And Clausius's entropy would first describe the perfectly controlled, totally reversible conversions of calorific and mechanical energies.

But we know that the story did not stop there, and the idea that uncontrolled transformations, sources of loss, always contribute to the irreversible increase in entropy was transformed into the affirmation of a growth: natural processes increase entropy. This is an example of the shifts that we mentioned: the interest in natural processes transforms the meaning of what was originally an engineering problem.

For the first time, it is not the manipulable that is subjected to analysis but rather that which, by definition, escapes manipulation or can only be subjected to it with ruses and losses. And thus physics recognized that dynamics—which describes nature as obedient and controllable in its being—only corresponds to a particular case. In thermodynamics, the controllable character is not natural but the product of artifice; the tendency to escape from domination manifests the intrinsic activity of nature. To nature, all states are not alike.

The nineteenth century, being at the same time haunted by the depletion of resources and carried away by the perspectives of revolution and progress, could not ignore irreversibility. And the twentieth century, in turn, sought in irreversible processes a key to what it sought to understand in nature, to those phenomena to which it had to give a physical status—under the threat of having to renounce the idea of the relevance of physical description in the understanding of nature. If the fear of depletion, the leveling of productive differences, was determinative for the original interpretation of the second principle, it is the biological model that constituted the decisive source of inspiration with respect to the history that followed. The refusal to restrict thermodynamics to systems artificially cut off from the world was produced by the desire to approach a world peopled by beings capable of evolving and innovating, of beings whose behavior we cannot render foreseeable and controllable except through enslaving them.

The thermodynamics of irreversible processes discovered that the fluxes that pass through certain physicochemical systems and keep them away from equilibrium can nourish phenomena of spontaneous self-organization, ruptures of symmetry, evolutions toward a growing complexity and diversity. There, where the general laws of thermodynamics stop, the constructive role of irreversibility can appear; it is the domain where collective behaviors are born and die, or transform themselves into a singular history that weaves together the uncertainty of fluctuations and the necessity of laws.

We are now closer to that nature which, according to the rare echoes that reach us, the pre-Socratics reflected on, and also to the sublunary nature whose powers of growth and corruption Aristotle described, to the inseparable intelligibility and incertitude of which he spoke. The paths of nature cannot be predicted with certainty; an element of chance is unavoidable and far more decisive than Aristotle himself realized: a bifurcating nature is one where small differences, insignificant fluctuations, can, if they occur in opportune circumstances, invade the whole system and create a new regime of functioning.

We also found this intrinsic instability of nature at another level—the microscopic. There, we were trying to understand what status to give to irreversibility, to the aleatory element, to statistical fluctuation, to all those notions that macroscopic science had just brought together; for, in the homogeneous world described by the usual laws of dynamics, or by any other system of laws of the same type, these notions would have only been approximations, and the perspectives that we have introduced, illusions.

The idea that physics cannot define molecular movement as determined, and thus that statistical description has an irreducible character, is nevertheless not unknown in physics. In particular, as the historian of science Brush has remarked,[1] some nineteenth-century men of science already spoke of indetermination, irregularity, the aleatory character of molecular movements, notably even then to justify the use of statistical reasoning. For example, Maxwell, in the article "Atom" published in 1875 in the *Encyclopedia Britannica*, wrote that the irregularity of elementary movement is necessary for the system to behave in an irreversible manner. But elsewhere he had affirmed that the irregularity was tied to our ignorance. In a general way, the ambiguity shifted between an intrinsic indetermination and an "epistemological" indetermination. This ambiguity, as we now know, was transformed into an opposition with the problem of interpretation of the quantum formalism.

Maxwell nevertheless caught a glimpse of the solution that we can today bring to this problem, when he spoke of the instability of movement, of singular points where small causes produce huge effects. But today, dynamics allows for the definition of systems where these singular points are literally everywhere, where no region of phase space, however small it may be, is deprived of them.

That being the case, the problem can be formulated in a general way. The ideal of omniscience is embodied in the science of trajectories, and in Laplace's demon, which contemplates them for an instant and calculates them for eternity. But the trajectories that appear so real are in fact idealizations: we never observe them as such because for that it would be necessary to have an observation of positively infinite precision. It would be necessary to be able to attribute to a dynamic system a perfectly precise initial condition, locating it in a unique state, to the exclusion of all other states, *however close they may be.* In the situations we normally think of, this remark is of no consequence. It is of little importance that the trajectory can only be defined approximately; the passage to the limit, toward the well-determined values of initial conditions, even if not effectively realizable, remains conceivable and the trajectory may be defined as the limit toward which tends a series, of growing precision, of our observations. However, we have met two types of insurmountable obstacles to this passage to the limit. These are the disorder, the chaos of trajectories for "unstable" systems and the coherence of the quantum behavior as determined by Planck's constant. In the first case, because divergent trajectories are found mixed together in the most intimate way, or, in the second, because they are "bound up" with each other, the definition of an exact state loses its sense; the trajectory is no longer only an idealization but an inadequate idealization.

Thus, dynamics and quantum mechanics have discovered the intrinsic limits of what was called the "scientific revolution," that is, the exceptional character of the situations that were the object of the initial experimental dialogue. The early physicists had very judiciously chosen objects that were clearly reducible to a mathetical modelization, objects that all belonged to a rather restricted class of dynamic systems whose trajectory could be meaningfully defined. The history of contemporary physics is tied to the discovery of the limited validity of the concepts developed for such systems, whose description could be presented as complete and deterministic, to the discovery, at the very heart of mathematical physics, of the "sublunar" world.

Of course, the end of the ideal of omniscience is the end of a problem posed solely at a theoretical level. No one has ever claimed to be capable of predicting the trajectories of a complex dynamic system. Laplace's demon itself appeared in the introduction to a treatise on probabilities. Laplace's demon was not the figure of universal mastery; it did not guarantee us the possibility of foreseeing everything; it stated that, from the point of view of physical theory, the future is contained in the present, while becoming and innovation, the world of processes where we live and that constitutes us, are, if not an illusion, at the very least appearances determined by our method of observation.

At both the macroscopic and microscopic levels, the sciences of nature are thus liberated from a narrow conception of objective reality, which believes that it must in principle deny novelty and diversity in the name of an unchanging universal law. They are freed from a fascination that represented rationality as closed and knowledge as in the process of completion. They are from now on open to unpredictability, no longer viewed in terms of an imperfect knowledge, or of insufficient control. Thus, they are open to a dialogue with a nature that cannot be dominated by a theoretical gaze, but must be explored, with an open world to which we belong, in whose construction we participate. This opening was well described by Serge Moscovici when he christened it the "Keplerian revolution," in opposition to the Copernican revolutions, which maintained the idea of an absolute point of view. Some texts have accused science and assimilated it to the undertaking of the disenchantment of the world. Let's quote Moscovici when he describes these sciences that are now being invented:

Science has become involved in this adventure, our adventure, in order to renew everthing it touches and warm all that it enters, the earth on which we live and the truths that enable us to live. At each turn it is not the echo of a demise, a bell tolling for a passing away, that is heard, but the voice of rebirth and beginning, ever afresh, of humanity and materiality, fixed for an instant in their ephemeral permanence. This is why the great discoveries are not revealed on a deathbed like that of Copernicus, but offered, like Kepler's, on the road of living dreams and passion.[2]

It remains for us to review some of the consequences of the metamorphosis of science whose history we have outlined.

Rediscovering Time

After more than three centuries, physics has rediscovered the multiplicity of times.

Einstein is often credited with the audacity of having envisaged time as a fourth dimension. But Lagrange, and also d'Alembert in the *Encyclopédie*, had already proposed that duration and the three spatial dimensions formed a unity of four dimensions. In fact, to affirm that time is nothing other than the geometrical parameter that allows calculation from the exterior, and as such, negates the becoming of all natural beings, has been almost a constant of the tradition of physics for the last three centuries. Thus, Émile Meyerson was able to describe the history of modern science as the progressive realization of what he regarded as a constitutive bias of human reason: the need for an explanation that reduces the diverse and the changing to the identical and the permanent, and as a result *eliminates time*.

In our era, it is Einstein who embodies with the greatest force the ambition of eliminating time. And he did this throughout all the criticisms, all the protests, all the distress stirred up by his absolute assertions. One scene is well known; it took place at the Société de Philosophie de Paris, on April 6, 1922.[3] Henri Bergson tried to defend, against Einstein, the multiplicity of lived times coexisting in the unity of real time, to argue for the intuitive evidence that makes us think that these multiple durations participate in the same world. Look at Einstein's response: he totally rejects, as incompetent, the "time of philosophers," convinced that no lived experience can save what science denies.

Perhaps even more remarkable is the exchange of letters between Einstein and the closest friend of his youth in Zurich, Michele Besso.[4] Besso was a scientist but, at the end of his life, was preoccupied more and more intensely by philosophy and literature, with everything that knits together the significance of human existence. As a result, he never stopped asking Einstein: What is irreversibility? What is its relation to the laws of physics? And Einstein replied, with a patience that he only showed to this particular friend: irreversibility is nothing but an illusion, created by improbable initial conditions. This dialogue, which led nowhere, continued until, in a final letter, at the death of Besso, Einstein wrote to his widow: "Michele has left this strange world a little ahead of me. This is of no importance. For us convinced physicists, the distinction between past, present, and future is only an illusion, however persistent."

Physics, today, no longer denies time. It recognizes the irreversible time of evolutions toward equilibrium, the rhythmic time of structures whose pulse is nourished by the world they are part of, the bifurcating time of evolutions generated by instability and amplification of fluctuations, and even microscopic time, which manifests the indetermination of microscopic physical evolutions. Every complex being is composed of a plurality of times, connected together by way of subtle and multiple articulations. The history, whether of a living being or of a society, will never again be able to be reduced to the monotonous simplicity of a unique time, whether this time expresses an invariance or traces the paths of progress or decline. The opposition between Carnot and Darwin has given way to a complementarity that we need to appreciate in each of its singular productions.

The discovery of the multiplicity of times is not a "revelation" that suddenly arises from science; quite to the contrary, men of science have now stopped denying what, so to speak, *all of us knew.* That is why the history of a science that denied time was also a history of social and cultural tensions.

What in the beginning had been an audacious wager against the dominant Aristotelian tradition first turned progressively into a dogmatic affirmation directed against all those—chemists, biologists, and physicians, for example—who sought to have the qualitative diversity of nature respected. But at the end of the nineteenth century, the confrontation was no longer there; it was no longer situated so much between scientists—from then on organized into differentiated academic disciplines—as between "science" and the rest of culture, in particular, philosophy. It is, moreover, possible to see in certain oppositions almost hierarchically established within the interior of the philosophical doctrines of this period evidence of confrontation with the dogmatism of scientific discourse. Thus, the "lived time" of phenomenologists, or the opposition between the objective world of science and the *Lebenswelt* that must elude it, could owe some of their characteristics to the necessity of establishing an ultimate bastion against the ravages of science. We have described the pretensions of science as linked to one of its historically and intellectually confined stages, but, for some, it was a question of ultimate stakes, involving the vocation or destiny of humankind, confrontations where the salvation or ruin of man was being played out. Thus, Gérard Granel reminds us that, according to Husserl, philosophy, a meditation on the originary entrenchment of all experience, is at war against an oblivion that would expose modern humanity to living, with all its sciences and efficiency, in the ruined monument of philosophy—which, for Husserl, *had made* the European world and its science—like the apes in the temple at Angkor.[5] There is a whole set of oppositions: between appearance and reality (with the question of who, science or philosophy, will be the judge), between knowledge and ignorance, between blind prejudice and knowledge produced by a rupture or asceticism, between the science of fundaments and the science of epiphenomena, which structure the site of a confrontation from which we would now like to distance ourselves as much as possible. In any case, as far as physicists are concerned, they have lost any *theoretical* argument that could claim any privilege, whether of extraterritoriality or of precedence. As scientists, they belong to a culture to which they in their turn contribute.

Actors and Spectators

Here again, it is perhaps Einstein who allows us to understand in the most dramatic way the meaning of the transformation undergone by physics during the course of this century. It is Einstein who first renewed the fecundity of *demonstrations of impossibility*, when he used the impossibility of transmitting information faster than the speed of light as the basis for excluding the notion of absolute simultaneity at a distance, and, on the basis of excluding this "inobservable," constructed the theory of relativity. Einstein saw this approach as equivalent to the one that based thermodynamics on the impossibility of perpetual motion. But certain of his contemporaries, like Heisenberg, clearly saw the significance of the difference between the two impossibilities; in the case of thermodynamics, a particular situation is defined as absent from nature; in the case of relativity, it is an observation that is defined as impossible, that is, a type of communication between nature and the one who describes it. And it is in following, in spite of Einstein, the example of Einstein that Heisenberg based the quantum formalism on the exclusion of magnitudes defined by physics as unobservable.

In his *Résumés de cours*, Merleau-Ponty had affirmed that the "philosophical" discoveries of science, its fundamental conceptual transformations, often come from *negative discoveries*, the occasion and starting point for a reversal of perspective.[6] The demonstrations of impossibility, whether they be in relativity, quantum mechanics, or dynamics, have taught us that nature cannot be described "from the exterior," as if one were an ideal, godlike spectator. Description is a communication, and this communication is subject to very general constraints that physics can learn to recognize inasmuch as these constraints identify us as macroscopic beings situated in the physical world. Physical theories from now on presuppose the definition of possibilities of communication with nature, the discovery of questions it cannot answer — unless it is we who cannot understand its eventual answers.

The very nature of the theoretical arguments with which we clarify the new position of physical descriptions manifests the double role, of actor and spectator, that is from now on assigned to us. Thus, even in the theory of dynamics of unstable systems, or in quantum mechanics, we continue to refer to notions of position in phase spaces and trajectories, which define us as ideal, godlike spectators, but this is precisely to indicate how in both cases it is a matter of inadequate idealizations. We thus maintain certain themes usually associated with "idealism," but it is quite remarkable that the most determinant requirements in the adoption of the new conceptual position that transform their meaning are usually those associated with "materialism": understanding nature in such a way that there is no absurdity in affirming that it produced us.

It is possible to situate our double role of actor and spectator in a context that clarifies the position of theoretical knowledge such as the evolution of physics now allows us to conceive of it. We would like to bring to light the coherent link that is now possible between what classical science had placed in opposition, namely, the disembodied observer and the object described from a position of overview. Of course, to go beyond this opposition, showing that from now on the concepts of physics contain a reference to the observer, in absolutely no way signifies that this observer must be characterized from a "biological," "psychological," or "philosophical" point of view. Physics is content to attribute to it the type of property that constitutes the necessary (but not in any way sufficient) condition for any experimental relation to nature, the distinction between past and future, but the demands of coherence lead one to ask if physics can also rediscover this type of property in the macroscopic world.

Let's start, for example, with this observer. As we have just said, the only thing that is required of it is an activity orientated in time, without which no exploration of the environment—and, a fortiori, no physical description, whether reversible or irreversible—is conceivable: the very definition of a measuring device, or the preparation of an experiment, necessitates the distinction between "before" and "after," and it is because we know the irreversibility of becoming that we are able to recognize the reversible movement, the simple change, reducible to a reversible equivalence between cause and effect. But classical dynamics constitutes in turn a point of departure; for the reversible laws of dynamics constitute *for us* the center of reference for the history of the mathematization of nature. The lawful world of reversible trajectories thus remains at the heart of our physics; it constitutes a conceptual and technical reference necessary for defining and describing the domain where instability allows the introduction of irreversibility, that is, a rupture of the symmetry of equations in relation to time. However, the reversible world is then no more than a particular case, and dynamics, equipped with the entropy operator that allows one to describe the complex world of processes, finds itself, in turn, taken as a point of departure: it can, at the macroscopic level, understand the monotonous inertia of states of equilibrium — average states produced by statistical compensation — but it can also predict the singularity of dissipative structures born far from equilibrium, and finally the simplest model for history, taken as a singular evolutionary path that crosses a succession of bifurcations. One can affirm, with respect to a structure formed from such an evolution, that its activity is the product of its history and therefore contains the distinction between past and future. Thus the circle is closed again; the macroscopic world is in turn capable of furnishing us with the starting point that we need for all observation. Summing up this circular schema:

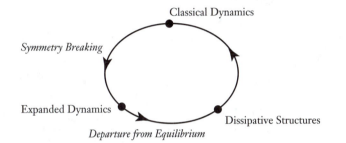

In opposition to the completely ideal reversibility of classical dynamics are two styles of becoming now corresponding to irreversible processes. The one, starting from an improbable past, heads for probable equilibrium; the other is open to a more properly historical future, that of dissipative structures that constitute the chance of aleatory fluctuations. But no logical necessity demanded that, in nature as understood by physics, dissipative structures actually exist; the "cosmological fact" of a universe capable of maintaining certain systems far from equilibrium was needed in order for the macroscopic world to be a world peopled with "observers," that is, a *nature*. Consequently, this schema does not express a truth of a logical or epistemological order, but the truth of our situation as macroscopic beings in a far-from-equilibrium world. It also expresses the historical truth of *our* physics, which was constituted with respect to the description of reversible and deterministic behaviors, and today no longer attributes to them the role of fundamental reality, but still that of a frame of reference. It appears to us essential that this schema not presume any fundamental mode or moment: each of the three modes is involved in the chain of implications, which expresses the new type of internal coherence that contemporary physics can lay claim to.

The schema that we have just described links descriptions that had each previously claimed preeminence. In a more general way, when it is a question of descriptions of the complex systems, living and social, that now concern us, it is clear that the preeminence of any neutral, godlike description is more than ever excluded, and that any theoretical model presupposes the choice of the question.

That is a lesson of wisdom that is important to underscore. Today, the so-called exact sciences need to get out of the laboratories where they have little by little learned the need to resist the fascination of a quest for the general truth of nature. They now know that idealized situations will not give them a universal key; therefore, they must finally become again "sciences of nature," confronted with the manifold richness that they have for so long given themselves the right to forget. From now on, they will be faced with the problem that some have wanted to reserve for the human sciences — whether it be to elevate or to diminish them — the necessary dialogue with preexisting knowledges concerning situations familiar to everyone. No more than the sciences of society can the sciences of nature forget the social and historical roots that create the familiarity necessary for the theoretical modelizing of a concrete situation. Thus, it is more necessary than ever not to make an obstacle of these roots, not to move from the relativity of our knowledges to some disenchanted relativism. In his reflection on the situation in sociology, Maurice Merleau-Ponty had already stressed this urgency, the urgency of thinking what he called a "truth in the situation."

So long as I keep before me the ideal of an absolute spectator, of knowledge in the absence of any viewpoint, I can only see in my situation a source of error. But once I have acknowledged that through it I am grafted onto all actions and all knowledge that can have a meaning for me, and that it gradually contains everthing that can be for me, then my contact with the social in the finitude of my situation is revealed to me as the origin of all truth, including that of science, and, since we have an idea of the truth, since we are inside truth and cannot get outside of it, all that I can do is define a truth within the situation.[7]

Thus science today affirms itself as *human* science, a science made by people for people. Within a rich and diverse population of cognitive practices, our science occupies the singular position of a poetic listening to nature — in the etymological sense that the poet is a maker — an active exploration, manipulating and calculating but now capable of respecting the nature that it makes speak. It is probable that this singularity will continue to arouse the hostility of those for whom any calculation or manipulation is suspect, but it is no longer the arrogant singularity that quite legitimately incited certain summary judgments about classical science.

A Whirlwind in a Turbulent Nature

Up to this point we have remained within a strictly scientific problematic. However, there is no reason to limit ourselves in this way; philosophy has always sought out ways of answering its questions, wherever they may be found, and, for its part, theoretical physics can now understand the sense of certain philosophical questions that relate to man's situation in the world. For example, we can comment on the dynamic transformation, from the model of stable systems, whose trajectories could be calculated, up to the discovery of instability, by way of a double philosophical reference: the Leibnizian monads and the Lucretian clinamen, two philosophical constructions among those that have been criticized as the most venturesome.[8] The clinamen, which disturbs "without reason" the trajectories of Lucretius's atoms, has often been dismissed as absurd and inconsequential; Leibniz's monads, metaphysical unities with no communication between them, "without windows through which something can get in or out," have been described as a "logical delirium."

Now, as we have seen, it is a property of any system whose trajectories are exactly calculable that one can give it a privileged "cyclic representation": in terms of entities without interaction, such that each one deploys for itself, as if alone in the world, a pseudoinertial movement. Each one of them then expresses, throughout all its movement, its own initial state, but this expression includes, as if in a preestablished harmony, its relations with all the others. In this representation, each state of each entity, as well as being self-determined, reflects at each moment the state of the whole system in its smallest details. This is no less than a definition of the Leibnizian monad. To go even further: a cursory way of describing the stationary states that constitute the electronic orbits of Bohr's atom is to say that they constitute so many monads.

Thus, we can express the physical property discovered by Hamiltonian dynamics in this form: all integrable systems (as they can by definition be represented in this cyclic representation) can be given a monadic representation. And, inversely, the Leibnizian monadology can be expressed in dynamic language: *the universe is an integrable system.*

Should we speak here of coincidence? Wouldn't the mathematical equivalence between the Newtonian representation, which requires masses and forces, and the monadic representation where each unity deploys in a spontaneous evolution the internal law of its behavior, be the expression, in the form of a physicomathematical property, of the fact that both of them are based on the same philosophical choice: the preeminence accorded to being over becoming, to permanence over change? Leibniz, the father of dynamics, certainly had not ignored what Whitehead emphasized:[9] the Newtonian forces only establish purely exterior relations between masses, which are nothing but their indifferent support; they are incapable of causing a becoming that is other than the eternal and monotonous repetition of an unchanging truth.

But the processes of absorption and emission of photons, the source of the first experimental facts that were at the basis of quantum mechanics, are sufficient in themselves to establish that this is not the whole story: they constitute, between the "monadic" electronic orbits, an interaction that no formal transformation can eliminate.

The physics of processes leads us to introduce a third representation, irreducible to the Leibnizian and Newtonian representations, which describes change neither in terms of well-defined unities that are autonomous and without interaction, nor entirely in terms of their interactions. The third representation describes real unities (photons, electrons), which, by definition, participate in dissipative processes that cannot be eliminated by transformation. These unities, contrary to the simple Newtonian "supports of forces," imply irreversible interaction with the world; their physical existence itself is defined by the becoming in which they participate.[10]

Without further developing these new perspectives, we propose, in order to recognize the convergence between theoretical physics and philosophical doctrine with respect to the articulation between being and becoming, to call this third representation the "Whiteheadian" representation. Whitehead wrote: "The elucidation of meaning involved in the phrase 'all things flow' is one chief task of metaphysics."[11] Today, physics and metaphysics meet to think a world where process and becoming would be constitutive of physical existence and where, contrary to the Leibnizian monads, the entities in existence could interact, and thus also be born and die.

Another philosophical interrogation that we can reread is that of dialectical materialism, and its search for universal laws to which the dialectical becoming of nature would respond. As for the materialists who wanted to conceive of a nature capable of history, the laws of reversible mechanics have been an obstacle for us, but we have not declared them false in the name of another type of universal law. Quite to the contrary, although we discovered the limits of their field of application, we preserved their fundamental character; they constitute the technical and conceptual reference that is necessary for the description and definition of the domain where they no longer suffice to characterize movement.

This role of reference to a lawful and ordered world, and, more technically, to the monadic theory of parallel evolutions, is precisely the role played by the fall (also parallel, lawful, and eternal) of the Lucretian atoms in the infinite void. We have already mentioned the clinamen and the instability of laminary flows. Here the possibility of an interpretation less tied to a particular physical phenomena presents itself. As Michel Serres has shown,[12] the infinite fall provides a *model* on which to base our conception of natural genesis, the disturbance that causes things to be born. Without the clinamen, which perturbs the vertical fall and leads to encounters, even associations, between the heretofore isolated atoms, each in its monotonous fall, no nature could be created, for the only thing perpetuated would be the connections between equivalent causes and effects governed by the laws of fate (*foedera fati*):

Denique si semper motus connectitur omnis et uetere exoritur (semper) nouus ordine certo nec declinando faciunt primordia motus principium quoddam quod fati foedera rumpat, ex infinito ne causam causa sequitur, libera per terras unde haec animantibus exstat...?[13]

Lucretius may be said to have *invented* the clinamen, in the same way that relics or archaeological treasures are invented: one "knows" that they are there before one begins to dig and actually discover them. And, in the same way, contemporary physics has invented irreversible time. Because, if only monotonous and reversible trajectories existed, where would the irreversible processes that we create and live come from? We "knew" that time was irreversible, and that is why the discovery of the instability of the trajectories of certain systems was a source of innovation, an opportunity grasped for an enlargement of dynamics.

Where trajectories cease to be determined, where the *foedera fati* governing the ordered and monotonous world of deterministic evolutions break down, nature begins. It also marks the beginning of a new science that describes the birth, proliferation, and death of natural beings. "The physics of falling, of repetition, of rigorous series is replaced by the creative science of chance and circumstances."[14] The *foedera fati* are replaced by the *foedera naturae*, which, as Serres emphasizes, designate both the "laws" of nature—local, singular, historical relations between things—and an "alliance," a contract with nature.

Thus, in Lucretian physics we again find the link we discovered within modern knowledge between the decisive choices underlying a physical description and a philosophical, ethical, or religious conception relating to man's situation in nature. The physics of universal connections is confronted with another science that no longer fights against disturbance or indetermination in the name of law and mastery. The classical science of fluids, from Archimedes to Clausius, was opposed to the science of turbulences and bifurcating evolutions, opposed to a science that shows that, far from the canals, disturbance can cause things, and nature, and man to be born.

It is here that Greek wisdom reaches one of its pinnacles. Where man is in the world, of the world, in matter, of matter, he is not a stranger, but a friend, a member of the family, a table companion, and an equal. He maintains a sensual, venereal pact with things. Conversely, many other wisdoms and many other sciences are based on breaking this pact. Man is a stranger to the world, to the dawn, to the sky, to things. He hates them and fights against them. His environment is a dangerous enemy to be fought and enslaved.... Epicurus and Lucretius live in a reconciled world, where the science of things and the science of man coincide. I am a disturbance, a whirlwind in a turbulent nature.[15]

An Open Science

We can also take up another type of rereading, focused this time on the mode of development particular to science. This internal dynamic of science can be described in terms of quite vast panoramas, of questions that continually reemerge, of slowly changing rhythms. There has been little real irreversibility in the history of science, few questions definitively abandoned or no longer valid. In its most classical description, the evolution of science has often been compared with the evolution of species: an arborescence of more and more diverse and specialized disciplines, an irreversible and unidirectional progress. We would like to shift from a biological to a geological image, because what we seek to describe is rather of the order of a drifting than a mutation. Questions that have been abandoned or repudiated by one discipline have moved silently into another, reappearing in a new theoretical context. It seems to us that their journey, whether underground or on the surface, manifests the silent work of questions that determine the deep communications beyond the proliferation of disciplines. And it is often at the intersections between disciplines, at the convergence between separate paths of approach, that the problems we thought were resolved reappear, that old questions predating the compartmentalization of disciplines have reemerged in a new form.

From this point of view, it is characteristic that many of the conceptual surprises that the evolution of the sciences has produced have the fatal charm of long-term revenge. The discovery of spectra in emission and absorption that led to the introduction of the notion of a quantum operator, and thus to the most decisive break with the classical science of trajectories, is in a way the revenge of the ancient chemists, who did not manage, during their time, to assert the specificity of chemical matter against the generality of interacting masses. At the intersection of dynamics and the science of chemical elements, the question they asked can no longer be ignored. And have not Stahl's vitalist claims also been avenged, since, at the fecund intersection between physicochemistry and biology from which molecular biology arises, one hears it affirmed that the only biological processes that physics can deduce from its laws are decomposition and death? We should not forget the revenge of those conquered by Newtonian science: the fatal announcement, at the center of the success of this science, of the mathematical law of heat diffusion that would forever make physicochemistry a science of processes, a science irreducible to classical dynamics.

The history of the sciences does not have the simplicity attributed to the biological evolution toward specialization; it is a slier, more subtle, and more surprising history. It is always capable of going backward, of rediscovering forgotten questions within an intellectually transformed landscape, of undoing the compartmentalization that it established, and, above all, of going beyond the most profoundly entrenched prejudices, even those that appear to be fundamental to it.

Such a description finds itself in clear contrast to the psychosocial analysis with which Thomas Kuhn has updated certain essential elements of the positivist conception of the evolution of the sciences: the evolution toward specialization and the growing division of scientific disciplines, the identification of "normal" scientific behavior with the "serious," "silent" researcher who wastes no time on "general" questions about the overall significance of his research and sticks to the specialized problems of his discipline; and the essential autonomy of scientific development from cultural, economic, and social problems.[16]

It is not our concern to question the validity of this description of scientific activity. In any case, it is sufficient to underline its partial and historically situated character. *Historically situated* means that scientific activity corresponds far better to Kuhn's description as carried on in the context of modern universities, where research and the initiation of future researchers are systematically associated within an academic structure that took shape throughout the nineteenth century, but that was previously nonexistent. It is in this structure that one finds the key to disciplinary implicit knowledge, to the "paradigm" that Kuhn posits as the basis of the normal research undertaken by a scientific community. It is by repeating, in the form of exercises, the solutions to the key problems solved by previous generations that students learn the theories on which research within a scientific community is based, but also the criteria that define a problem as interesting and a solution as acceptable. The transition from student to researcher indeed takes place, in this type of teaching, without discontinuity: the researcher continues to solve problems that are identified as essentially similar to the model problems, applying similar techniques to them; it is just a question of problems that no one had previously solved. *Partial* means that, even in our era, for which Kuhn's description has the greatest degree of relevance, it only concerns at best one dimension of scientific activity, more or less important according to individual researchers and the institutional context in which they work.

We can best clarify this remark by considering Kuhn's conception of paradigmatic change. This transformation will often appear as a *crisis*: instead of remaining a silent, almost invisible norm, instead of being "taken for granted," the paradigm is discussed and questioned. Instead of working in unison toward the resolution of generally accepted problems, the members of the community ask fundamental questions, challenging the legitimacy of their methods. The group, rendered homogeneous by its education concerning research activity, begins to diversify, and different points of view, cultural experiences, and philosophical convictions are expressed and often play a decisive role in the discovery of a new paradigm. Its appearance further increases the intensity of the discussion. The rival paradigms' respective domains of fecundity are put to the test until a difference, amplified and stabilized by academic circuits, decides the victory of one of them. Little by little, with the new generation of scientists, silence and unanimity reestablish themselves, new textbooks are written, and once again things "go without saying."

In this view, the driving force of scientific innovation is clearly the extremely conservative behavior of scientific communities that stubbornly apply to nature the same techniques, the same concepts, and always end up encountering an equally stubborn resistance from nature: nature refuses to express itself in a language that presupposes paradigmatic rules, and the crisis we have just described explodes with all the more force in that it results from blind confidence. From then on, all intellectual resources are concentrated on the search for a new language dealing with a set of problems now considered decisive: namely, those that have elicited the resistance of nature. Thus, scientific communities systematically provoke crises, but to the extent that they are not looking for them.

The questions that we have chosen to investigate in the history of the sciences have led us to explore very different dimensions from the ones that interest Kuhn. We have above all dwelled on the continuities, not the "obvious" continuities, but the more hidden ones, about which certain scientists have never ceased asking questions. It seems to us that one should not seek to understand why there is continuity, from generation to generation, in the debate on the specificity of complex behaviors, on the irreducibility of the science of fire and transformations of matter to the description of masses and trajectories; for us, it is more a question of knowing how such problems, the problems of Stahl, Diderot, or Venel, could have been forgotten.

For the past century, the history of physics has obviously shown us crises that resemble Kuhn's descriptions, crises that scientists experience without having sought them out, crises that philosophical preoccupations may well have triggered, but only in a situation of instability determined by the fruitless effort of extending a paradigm to certain natural phenomena. But it also shows us lineages of problems clearly and deliberately created by philosophical preoccupations. And it establishes the fecundity of such an approach. Scientists are not doomed to behave like Kuhnian sleepwalkers; they can, without having to give up being scientists, take the initiative, seeking to integrate new perspectives and questions into the sciences.

The history of the sciences, like all social histories, is a complex process in which events determined by local interactions coexist with projects informed by global conceptions about the task of science and the aim of knowledge. It is also a dramatic history of ruined ambitions, failed ideas, and accomplishments that do not achieve the significance that they should have attained. Once again, Einstein can serve as an example: with relativity, the quantification of energy, and the cosmological model, he dealt the first blows against the classical conception of the world and knowledge, even though his project was always a return to a universal, complete, and deterministic description of the physical world.

The drama of Einstein lies in this uncontrollable gap between the individual intentions of actors and the actual significance that the globlal context lends to their actions.

Scientific Interrogation

We have argued that the fundamentally open character of science be recognized, and that, in particular, the value of communications between philosophical and scientific interrogations cease being suppressed by compartmentalization or destroyed by a confrontational attitude. The philosophical "ratification" of the pretensions of classical science allows certain philosophers to situate and freeze the scientific approach, and from then on to give themselves the right to ignore it. This strategy has been dominant for a long time despite protests like those of Maurice Merleau-

Ponty, who wrote what, from a certain point of view, could constitute the best definition of the themes and objectives of this essay:

The recourse to science has no need to be justified: whatever conception one has of philosophy, it elucidates experience, and science is a sector of our experience.... it is impossible to impugn it in advance under the pretext that it works on the level of certain ontological presuppositions: if there are presuppositions, science itself, in its wanderings through being, will certainly find the occasion to challenge them. Being forces its way through science as it does through all individual life. By interrogating science, philosophy will gain an encounter with certain articulations of being that would be otherwise difficult for it to discover.[17]

But if no privilege, no precedence, no definitively fixed limit settles in a stable manner the difference between scientific and philosophical interrogations, it is nevertheless not a question of identity or their reciprocal substitution. We think that it is a question of the complementarity of knowledges that, in both cases, constitute the expression, according to more or less rigorous rules, of preoccupations belonging to a culture and an era. The question is thus one of rules, methods, and constraints.

We have explored some of the constraints to which scientific interrogation is subjected. On the one hand, experimental dialogue itself limits the freedom of scientists; they do not do what they want, and nature refutes their most seductive hypotheses, their most profound theories—from which comes, among other things, science's slow rhythm with respect to conceptual exploration, and the forever present temptation to extrapolate to the extreme the rare and limited "yeses" that have been obtained from nature. For example, the "triumph" of the science of trajectories was actually restricted by a problem as simple as that of the three bodies. On the other hand, a second constraint, as fruitful as the first but more recently brought to light, is the prohibition against basing a theory on magnitudes defined as unobservable in principle. Now, that is an interesting turnaround. Scientific objectivity had for a long time been defined as the absence of reference to an observer; now it finds itself defined by the condition of a meaningful "observational" relation—a reference to humankind, or to bacteria, for example, that other inhabitant of the macroscopic world whose movements truly constitute an exploratory activity since they presume an orientation in time and the capacity to react irreversibly to chemical modifications of the milieu. Our science, for a long time defined by research from a position of absolute overview, finally discovers itself as a "centered" science, the descriptions that it produces are situated and express our situation within a physical world.

It is possible that the situation appears rather different for philosophy. We would like here to attempt an assessment and risk a hypothesis. We have found inspiration from a certain number of philosophers, among them some who belong to our era, such as Michel Serres or Gilles Deleuze, or to the history of philosophy, such as Lucretius, Leibniz, and Whitehead. We have no intention of proceeding to some kind of amalgamation, but it seems to us that a trait common to those who have helped us to think through the conceptual metamorphosis of science and its implications is the attempt to speak of the world without passing through the Kantian tribunal, without putting the human subject defined by his or her intellectual categories at the center of their system, without subjecting their remarks to the criteria of what such a subject can, legitimately, think. In short, it is a question of precritical or acritical speculative thinkers.

How is it that we have found inspiration from speculative philosophers when reflecting on the discovery by physics of its open character? The hypothesis that we would like to offer is the following: for these philosophers, it is likewise a matter of an *experimental* approach—not an experimentation on nature but on concepts and their articulations, an experimentation in the art of posing problems and of following the consequences with the most extreme rigor.

Whitehead clearly expressed this conception of philosophical experimentation, with its own degree of freedom but also with its own constraints. Thus, he maintained that philosophy cannot have recourse to the strategy that underlies the experimental dialogue of science with nature—the strategy of choosing what is interesting and what can be neglected: "Philosophy destroys its usefulness when it indulges in brilliant feats of explaining away."[18]

One sees that, in our hypothesis, scientific and philosophical experimentation must not be put in opposition, as one would oppose the concrete and the abstract. Whitehead even inverted the opposition, reserving for philosophy the task of producing, through the play of quite abstract concepts, real experiences in their concrete richness. And Deleuze even goes so far as to speak, with respect to such a philosophical ambition, of *empiricism*.

Empiricism is by no means a reaction against concepts, nor a simple appeal to lived experience. On the contrary, it undertakes the most insane creation of concepts ever seen or heard. Empiricism is a mysticism and a mathematicism of concepts, but precisely one which treats the concept as object of an encounter, as a here-and-now, or rather as an Erewhon (N.B.: a utopia, and thus both "here-and-now" and "nowhere," imagined by Samuel Butler), from which inexhaustibly emerge ever new, differently distributed "heres" and "nows." Only an empiricist could say: concepts are indeed things, but things in their free and wild state, beyond "anthropological predicates." I make, remake and unmake my concepts along a moving horizon, from an always decentred centre, from an always displaced periphery which displaces and differentiates them.[19]

Erewhon, the unobservable par excellence from which arise the heres and nows, the multiplicity of real experiences — this is indeed an unfamiliar thought for those of us who have made the exclusion of what is unobservable in principle the resource of a new invention. And yet, it is exactly in thinking the unobservable, monads, clinamen, and eternal objects that, in certain cases, philosophers have "preceded" science, have explored concepts and their implications well before science could make use of them or discover their constraining power. It is without doubt here that resides the price of the risk accepted by those who do not restrict themselves to using the powers of the imagination in a heuristic manner, in order to inspire experimental and theoretical hypotheses, but carry them to their highest intensity with the strict demands of coherence and precision.

Here again, we must emphasize a convergence that reveals the cultural coherence of an era. The philosophers we have cited have given us the means, to use Deleuze's expression, of passing "from science to dream and back again," because they have been led by "an imagination which traverses domains, orders and levels, knocking down partitions, an imagination coextensive with the world, guiding our body and inspiring our souls, grasping the unity of nature and mind."[20] But, conversely, Deleuze calls on nature and the sciences of nature to describe the powers of the imagination and avoid any reference to the man of traditional philosophy, the active subject, endowed with projects, intentions, and will. "The Idea," he writes, "turns us into larvae, having put aside the identity of the I along with the resemblance of the self."[21] When seeking to understand the "dramatization," the terrible movement endured by whoever is preyed on by an idea, in whom an idea embodies itself, we should think of a larva, capable (contrary to the constituted organism, engaged in a stable activity) of undergoing terrible movements, lines, slippages, rotations; we need to think of those processes that the sciences of nature attempt to describe. "Dramatisation takes place under the critical eye of the savant as much as it does in the head of the dreamer."[22] The psychological dramatization finds its echoes in the geological, geographical, biological, and ecological processes that create spaces, model, and drastically alter landscapes, thereby determining the migrations, competitions, or mutual amplifications between processes of growth, proliferations, slow erosions, and brutal disintegrations.

The Metamorphoses of Nature

The metamorphosis of the contemporary sciences is not a rupture. On the contrary, we believe that it helps us to understand the significance and intelligence of knowledges and ancient practices that modern science, based on the model of automated technical production, believed it could ignore. Thus, Michel Serres has often evoked the respect that peasants and seafarers have for the world in which they live. They know that one has no control over time and that one cannot rush the growth of the living, the process of autonomous transformation that the Greeks called *physis*. In this sense, our science is at last on the way to becoming a physical science since it has to finally accept the autonomy of things, *and not only of living things*. Human activity contributes to the production of a new state of nature. As with the development of plants, the development of this new nature, peopled by machines and technology, the development of social and cultural practices, and the growth of cities are continuous and autonomous processes in which one can certainly intervene to modify or organize them, but whose intrinsic time must be taken into account, under threat of failure.[23] The problem posed by the interaction of human populations and machine populations has nothing in common with the relatively simple and controllable problem of the construction of this or that machine. The technological world that classical science contributed to creating needs quite different concepts from those of classical science in order to be understood.

When we begin to understand nature in the sense of *physis*, we can also begin to understand the complexity of the questions that confront the sciences of society. When we learn the "respect" that physical theory imposes on us with regard to nature, we must also learn to respect other intellectual approaches, whether they be the traditional approaches of seafarers and peasants, or approaches created by other sciences. We must learn no longer to judge the population of knowledges, practices, and cultures produced by human societies, but to interbreed with them, establishing novel communications that enable us to deal with the unprecedented demands of our era.

What is this world in relation to which we have learned the necessity of respect? We have evoked the classical conception of the world and the evolving world of the nineteenth century. In both cases, it was a question of control and of the dualism that opposes the controller and the controlled, the dominant and the dominated. Whether nature is a clock or a motor, or the path of a progress that leads toward us, it constitutes a stable reality of which we can be assured. What can we say of our world that has nourished the contemporary metamorphosis of science? It is a world that we can understand as natural in the very moment we understand that we are part of it, but in which the former certitudes suddenly disappear: whether it is a question of music, painting, literature, or morals, no model can any longer claim sole legitimacy; none is any longer exclusive. Everywhere, we see multiple experimentation, more or less risky, ephemeral, or successful.

This world that seems to have renounced the security of stable, permanent norms is clearly a dangerous and uncertain world. It can inspire no blind confidence in us, but perhaps the feeling of mitigated hope that certain Talmudic texts have, it seems, attributed to the God of Genesis:

Twenty-six attempts have preceded the present genesis, and all have been doomed to failure. The world of man has arisen out of the chaotic womb of the preceding debris, but it has no guarantee certificate: it too is exposed to the risk of failure and the return to nothing. "Let's hope it works" (Halway Sheyaamod), exclaimed God as he created the world, and this hope accompanies the subsequent history of the world and humanity, emphasizing right from the start that this history is stamped with the mark of radical insecurity.[24]

It is this cultural climate that nourishes and amplifies the discovery of undreamed-of objects, quasars with formidable energies, fascinating black holes, the discovery also, on earth, of the diversity of experiences that nature effects, theoretical discoveries, and finally the problems of instabilities, proliferations, migrations, and structurations. At the point where science had shown us an unchanging and pacified nature, we understand that no organization or stability is guaranteed or legitimate, that none can impose itself by right, that they are all the products of circumstance and at the mercy of circumstance.

That being the case, Jacques Monod was right: the old animist alliance is truly dead, and with it all the others that appeared to us as intentional, conscious subjects, endowed with projects, closed in on a stable identity and well-established habits, citizens at the center of a world made for us. The finalized, static, and harmonious world that the Copernican revolution destroyed when it launched the earth into infinite space, is definitely dead. But then, neither is our world that of the "modern alliance." It is not the silent and monotonous world, abandoned by the old enchantments, the clock world over which we received jurisdiction. Nature is not made for us, and it has not surrendered to our will. The time has come, as Jacques Monod informed us, to assume the risks of humankind's adventure, but if we can do it, it is because this is now the form of our participation in the cultural and natural becoming; this is the lesson expressed by nature when we listen to it. Scientific knowledge, drawn from the dreams of an inspired, that is, supernatural revelation, can today be discovered both as a "poetic listening" to nature and to natural processes in nature, open processes of production and invention, in an open, productive, and inventive world. The time has come for new alliances, which have always existed but for a long time have been ignored, between the history of humankind, its societies, its knowledges, and the exploratory adventure of nature.

F O U R

Turtles All the Way Down

ONE DAY when the philosopher William James, who had a liking for scientific popularization, had just finished explaining in a small American town how the earth revolved around the sun, he saw, according to the anecdote, an elderly lady approaching him with a determined look. Apparently, she strongly disagreed, expressing herself in the following terms: no, the earth does not move, because, as is well known, it sits on the back of a turtle. James decided to be polite and asked what, according to this hypothesis, the turtle rested on. The old lady replied without hesitating: "But on another turtle, of course." And James persisted: "But what does the second turtle rest on?" Then, so the story goes, the old lady triumphantly exclaimed: "It's no use, Mr. James, it's turtles all the way down."

After the initial amusement, the reply demands some consideration. At what stage are we? Are we, even today, able to tackle the epistemological "obstacle," the description of a world centered on us, of turtles made for our support, while wondering about the effects of analysis and absurdity produced from a traditional knowledge by that most potent weapon of our intellectual tradition, retaliation. What do we know and where are we, in this anecdote?

One thing is clear: William James is also conveying a foundational knowledge, the description of the solar system as it is said to exist objectively. And, starting from that point, the question has to be asked: is there much difference between the old lady's turtles and the fundamental laws of physics, if from these laws a physicist can claim that the totality of phenomena can in principle be understood? It may be argued that we are no longer in the period when Laplace's demon was able to deduce from the laws of dynamics and the instantaneous state of the world the totality of its past and its future. Are we, however, that far from it? The examples multiply indicating that we have not yet renounced the quest for a knowledge that, in one way or another, gives back to the world the transparency that classical reason had postulated, even if this entails producing the fantastic anachronism of the mind-matter dualism. Anything, for instance, seems preferable to what, since Niels Bohr, quantum mechanics asks us to think—that our "objective" description of physical phenomena always implies a communication or the possibility of communication, and that, like it, our description has two irreducible relational terms: the processes that we interrogate experimentally and the world of our practical and theoretical instruments. Anything to escape the idea that the physicist does not discover the world "as it is"; that the elementary phenomena it describes contain an intrinsic reference to the macroscopic world, to the world of our interactions and measurements; and thus, that when we speak of the laws of physics as if they gave us access to a fundamental reality, we produce the same paradoxical movement as the old lady with the turtles. We produce a grounding effect, a foundational effect starting from a reality that refers to us and that puts us at the center of our description.

So, we have just produced a chiasmus. We can no longer claim possession of an objective knowledge of the solar system because, today, the old lady's absurdity returns us to the concrete historical situation of physics; and it is the certitude of being able to describe the world as it is, independently of all observation, that we can recognize as the obstacle, whose fascination has kept us prisoners.

I wanted to tell the anecdote of the turtles so that these slow and obstinate, prehistoric-looking creatures might remind us of how much we are today the unaware prisoners of a few powerfully formalized languages. The world of the objects of our formalizations is so depopulated that the instruments of exploration transform themselves into a screening machine thanks to which the real is judged, the objective and the illusory separated. Perhaps the bestiary of our mathematical objects will become sufficiently dense—topological creatures, strange attractors, fractals, catastrophes: we are already becoming familiar with some of them—that we will be able to look back with amusement at the old turtle, at the class of languages that presume the world as the object of an ideally omniscient language. Today we are still coming to terms—evidenced by the resurgence of bizarre interpretations of quantum mechanics—with the limits of these languages through the shock of paradox, by the absurdity, assumed or refused, of a foundation that makes reference to that which it is supposed to ground.

A dangerous, unstable situation. All the more dangerous and unstable in that certain scientists feel the pressure of expectations. This symposium at Cerisy on the topic "self-organization from physics to politics" shows that we are asking a lot from science these days.[1] Some are demanding that it take itself in hand and justify a kind of reconciliation with mystical traditions; we see contemporary physics including consciousness in the very definition of the world; more and more frequently, physicists enter into dialogue with theologians to deride materialism, always described as naive. Others ask it to produce a "good" knowledge, a redeeming knowledge that, by its virtues alone, will reconcile us with the world. This is, for example, manifested by the sudden seduction of systems thinking. Everything that classical science had rejected became, by the waving of a magic wand, so many primary properties or explanatory principles. For example, self-preservation appears like a foundational tautology: a system would not exist, we are told, if it did not have as a goal the maintenance of its own existence. The real now seems "made to" produce the reassuring and meaningful organization of a both stable and creative world, where each part realizes itself by participating in the greater complexity of the next level, where we rediscover the ordered, reassuring, and yet open harmony that we apparently hope for. It is, I think, one of the principal stakes of this symposium to contribute to the disconnection of the ensemble of metaphorical circulations thanks to which, particularly, one sees political projects seeking a justification in the truth of a nature that science is supposed to discover. Whether this science be physics or biology, whether it is a question of self-organization or sociobiology, is of little importance; only the authority that we attribute to the knowledge in question, and the expectation that it will give us a law whose obedience will save us, are important.

One of the surest signs that social expectations both authorize and provoke the scientist is the proliferation of generalizations. We know the massive generalization on which sociobiology is based: that the whole of biological history can be reduced to selective constraints in such a way that every trait, every behavior will find its raison d'être in the optimization of an adaptive performance. As for myself, I can attest to the insistence with which the physicists and chemists at Brussels are asked to surrender to the temptation to produce a general theory of the self-organization of the physical and social world. Some kind of heroism is truly demanded from the scientist in order to resist such an insistence.

What, then, should be done? I do not think we should, in the name of the demands of purity, hunt down metaphorical communications or reestablish the separation between domains. This would be to defend not a hypothetical purity, but a fait accompli, since metaphorical circulation has already constituted the objects that we encounter in these domains. For example, the metaphors of society and organism have undergone such a dense circulation that it is pointless to speak of biologism in sociology, or of social projection in biology. Perhaps we know even still less about what a body is independent of the social than what a society is independent of biological organization. The real problem is not purity, but prestige and authority. Here is a theoretical position that can and should be resolutely defended: faced with the consequences and implications of their work, scientists are just like the rest of us; there is no point in expecting from them a particular lucidity concerning the scope and stakes of this work. In the same way that we accept that this work takes part, in the best sense of the term, in the questions and interests of its era, we have to accept that, with respect to the interpretations that scientists produce, they are prey to the same anxieties, the same temptations, the same expectations as their contemporaries. Einstein said: do not listen to the scientist saying what he or she does, look at what he or she does. We should not understand this warning in the sense of a distinction à la Bachelard between day work and nocturnal reverie, but draw the consequences from the fact that no privileged access of a creator to his work exists that can constrain us from judging positively or negatively what he or she does according to what he or she says about it.

But, if we cannot trust the authority of those who produce knowledge concerning the very knowledge they produce, where are we going to find the tools of evaluation? How are we going to orient ourselves? The response is probably none other than political and cultural. The question may be expressed like this: how are we going to re-create a culture, a sufficiently dense and critical social milieu of discussion and negotiation such that the stakes of the theoretical discourses are situated in their cultural and historical relativity and that the scientist cannot be, nor feel required to take on the role of, a prophet. As far as I am concerned, I have tried to find in the history of the problems a necessary counterbalance to the authority of the pretensions of contemporary knowledge — not to reduce them to nothing, but in order to situate them and measure their scope. This is a risky job since it looks like a partner to the sadly notorious search for precursors — a task that can find no guarantee in any method, and that must only seek such a guarantee in the critical confrontation with other interrogations. Once again, there is no point in expecting an individual solution, a solution that does not pass through the construction of a culture.

How do we define that which could claim, according to my way of thinking, a general cultural significance in the thermodynamic and kinetic theories that deal with what have been called the phenomena of self-organization? I think that far more than positive results, it is a question of freedom from a certain number of expectations, ideas, and a priori judgments that were shared not only by physicists but by many others who, often without realizing it, had accepted the theories of physics as a model of scientificity. What we can now hope for is a proliferation of gradually articulated, local theoretical languages, that is, a world in which theoretical turtles would no longer be an object of scandal or quest, in which we would no longer ask what they rested on but with whom they lived, in which we would recognize them as a rigorous exploration of problems accepted as both relative and fully positive.

Since its origin, physics has been dominated by the quest for general laws; it has also assumed that there is an identity between knowledge and the possibility of manipulating; finally, it has taken as its privileged object the state, which implies the conviction that the relevant description of a system can always be reduced to the definition of an instantaneous state. The notion of state function in dynamics, as in thermodynamics, epitomizes this triple anticipation, of generality, manipulability, and instantaneity. This is why, to put it bluntly, the conceptual transformation that I am going to discuss results from an awareness that, except in exceptional cases, the physics of processes cannot be reduced to a physics of states.[2]

In a way, the distinction between state and process is a very old object of debate similar to the arguments invoked by some eighteenth-century chemists to defend the specificities of their object against the abstract generalizations of physicists, as exemplified by Venel's article "Chymie" in the *Encyclopédie*. In this sense, it is possible to see in the contemporary transformation of physics the return of a problematic that belonged to chemistry before it was led to abandon its theoretical questions in order to dedicate itself to the analysis and synthesis of bodies. In this perspective, one should see the development of kinetic concepts within theoretical physics—from Planck's work on blackbodies up to current cosmological theories—as the progressive beleaguerment of this science by the long-suppressed and denied questions of chemistry. We should never forget that the present evolution of physics is partly at least a backward movement that has been endured rather than sought for.

The birth of thermodynamics belongs to this endured movement. Here, for the first time, a general description shows its powerlessness and acknowledged lack of relevance as far as the physical understanding of its object is concerned.

In short, at the end of its formalization process during the course of the eighteenth century, dynamics found itself capable of exhausting its object with a series of equivalences that define equally and inseparably the possibilities of work and manipulation. Any acceleration undergone by a body is defined and measured through the work that this acceleration allows the body to provide. Thus, the fact that a falling body *gains* speed by *losing* altitude, can be *invested* in any work equivalent to the work necessary to make the body regain its initial altitude. Now, when it becomes a question of no longer putting to work bodies subjected to dynamic forces but rather of putting physicochemical processes to work, one realizes that, if all those processes conserve energy, energy no longer allows one to differentiate between what is possible and what is not. Energy is no longer the adequate state function leading to the definition of physical evolution as a "change of state."

I will not spend too much time on the history of the introduction of the new state function that was called entropy, except to note that entropy itself is also tied to the problematic of work. But this time, it is no longer a question of putting to work an inert body but rather an active milieu whose activity will be thought of as essentially resistant to this operation. The physicochemical process par excellence, which received its mathematical description during the early years of the nineteenth century, is the diffusion of heat. And this occurs spontaneously, without it being possible to identify a manipulable relation of equivalence between cause and effect. From the point of view of classical reason, it is a scandal, which Émile Meyerson aptly called an *irrationality*.

It is we who seek to establish identity in nature, who bring identity to nature, who suppose that it can be found in nature. . . . This is what we call understanding nature or explaining it. In some ways nature lends itself to this, but it also protects itself from it. Reality rises up in revolt, no longer allowing us to suppress it. Carnot's law is the expression of the resistance that nature opposes to the constraints that our understanding, through the principle of causality, tries to exercise on it.[3]

And every time that a process escapes the active channeling whereby we both measure and utilize the transformations of a system, possibilities for work are lost, *dissipated*. Physicochemical processes will be globally described as dissipative.

Within physics the problematic of work, that is, the search for general relations of equivalence between what has been invested in a system and what it is capable of restoring, has had paradoxical consequences; thermodynamics was constituted in relation to irreversible processes but also *against* them, seeking not to know but to control them. This is probably why there was not an intellectual scandal when thermodynamics concentrated on equilibrium states, that is, states where processes are no longer possible, the state to which processes fatally lead if they occur in an isolated system. And the thermodynamics of equilibrium states again finds manipulable equivalences between cause and effect: a system whose state of equilibrium has been displaced with sufficient precaution such that no irreversible process can occur is ideally capable of restoring to the world the work that was invested to bring about the displacement in question. The theories of thermodynamic equilibrium are rational in Meyerson's sense.

Here, then, are the two privileged objects of classical physics, the dynamic object wholly intelligible in terms of reversible equivalences and the thermodymanic object with its privileged state, the state of equilibrium. The first is characterized by an absolute memory; nothing happens to it, nothing is produced that is not already contained in all of its previous states. The second, on the contrary, is characterized by an evolution during which its initial situation is forgotten, and at the end of which its properties are the general properties deducible from the thermodynamic state function. From this point of view, whether we are dealing with the ordered structure of a crystal, a gas, or a liquid, all thermodynamic states of equilibrium are equivalent; they all correspond to the disappearance of any process, to the oblivion of all singularity.

Summarized in this way, the story seems odd. The choices made by physics appear so partial and subjected to an a priori ideal that one cannot understand the authority they have been able to assume. And yet the physics of the twentieth century has been dominated by distinctions based on these choices between the real, the clearly established phenomenon subjected to the law of large numbers, and the illusory.

The embodiment of this distinction is none other than Maxwell's demon. Indeed, this demon is defined as dealing with the real beyond our phenomenological approach, that is, for instance, with the molecules of a gas. Armed with its racket, it can manipulate them. From its point of view, the evolution of a gas toward equilibrium does not appear as a law but as the most probable effect of the movement of molecules. But the demon who deals with molecules in such a way can decide to escape probabilities, to oppose the leveling of differences, and to create new ones within the general constraints that define dynamics. The evolution toward equilibrium is thus no more than a well-established phenomenon, a plausible anticipation in virtue of the laws of motion, but not a law. And the demon that deals with laws, deals by the same stroke with a world that can be fully put to work. For such a demon, all the distinctions between available energy and degraded energy are obliterated. As for illusion, it will be denounced in all the other cases, each time that which Meyerson called irrationality cannot be reduced to a probable leveling, to a progressive oblivion.

The discrimination carried out between the real, the probable, and the illusory indicates the site where the social relations of prestige and authority within science have the greatest bearing. The social debate is not primarily concerned with the contents of this or that theory but with the legitimacy of the problem that the theory presents, not with the discourse but with the metadiscourse defining what is and what is not acceptable for scientific discourse. And the possibility for some physicists to be recognized as qualified a priori to judge the real in terms of their conceptual instruments does not present in the first instance a psychological or philosophical problem—here we need to clearly separate our position from analyses like that of Meyerson—but a political problem. In particular, it is social isolation that turns the physicist into a prophet or a subservient technician, which leads him to deny with authority or contempt the concrete reality that he has no way of knowing. Consequently, the real question of physics is a political and social one. With respect to the new theoretical questions that reveal the partial and globally illegitimate character of former generalizations, we can only hope and work for them to accentuate the contradiction between the particularities of scientific research and the kind of socialization that has been ours since the nineteenth century.

Let's have a closer look at some of these generalizations. One of them prevails not only in physics but in all the sciences that deal with large populations constituted from elements that are essentially independent of one another. It is the conviction that the law of averages always applies, that what is called the law of large numbers establishes a strict distinction between individual behaviors and their statistical resultant. Now, this conviction, as I will show, relates quite precisely to the opposition that interests me between a physics of states and a physics of processes, because the law of large numbers as it was used by Boltzmann in the statistical interpretation of evolution toward equilibrium, presupposes quite specifically that the problem of processes is not dealt with. One considers the different instantaneous molecular configurations that are possible within a system without asking how they are going to be produced. And it is not necessary to ask this question, because one attributes to each configuration an a priori equal probability, and contents oneself with counting, that is, with evaluating the probability of different macroscopic states in terms of the number of different molecular configurations that each one of them realizes. In this interpretation, the state of equilibrium is the privileged state exactly insofar as it is the state in which processes, of whatever kind, are henceforth of no consequence; the overwhelming majority of a priori equally possible transformations move the system between two molecular configurations, both of which realize the state of equilibrium.

One finds, for all that, the same limitation in theories of information, which are also based on the a priori evaluation of arrangements or configurations of elements that are yielded by each instantaneous global arrangement. Here again, organization is seen as just an improbable global arrangement, and it is, for example, difficult to establish a distinction between the statistical order of a crystal and the production of periodic structures from hydrodynamic or chemical processes. In every case, from the point of view of information, there is redundance, because the description of the instantaneous elementary configuration of a small region of the system allows an extrapolation to a description of the whole system. And yet, the one, isolated, maintains indefinitely this redundance, while the other resembles Lewis Carroll's red queen; "it takes all the dissipation you can pay for to stay in the same regime."

On the contrary, the distinctive feature of kinetic models, to which theories like that of dissipative structures are directly related, is to attribute importance to processes as such. The problem posed by Boltzmann is thus reversed. It is no longer a question of calculating the probability of an instantaneous global state from the calculation of the number of configurations, as if the probability of these different instantaneous configurations were independent of the processes that produced them. On the contrary, kinetic models calculate the speeds of different processes as a function of the probabilities of the events that are capable of occurring in the system, and make these probabilities themselves the product of the evolution of the system. Thus, in the case of the phase transitions of equilibrium, it is necessary to take into account the fact that, if a droplet is formed in a gas, the chances that new molecules aggregate and that the drop develops rather than evaporates depends on the size that has already been attained, because the intensity of the forces of attraction increases with the size of the drop. As long as the molecules were separated, these forces were negligible, and one could characterize the different arrangements as if they were formed from essentially independent elements. But they play a decisive role when surrounding a droplet and determine the transitory character or the amplification of the process of aggregation. What could, in a gas at equilibrium, be omitted in the a priori calculation of probabilities becomes of decisive importance at the moment of phase transitions.

The common feature of kinetic models is to calculate not the probability of an instantaneous state but the probability of a history. The probabilities of instantaneous states are no longer evaluated a priori; instead, one will ask, for example, taking into account the processes at work in a system, what the probabilities are that a local event will entail consequences, be propagated, or be leveled out. This being the case, kinetic models allow new questions to be asked, notably one that has always fascinated the imagination, the problem of phase transitions, because inorganic nature experiences not only progressive evolutions toward disorder and indifference but also abrupt metamorphoses, discontinuous transformations: crystallization (order emerges from disorder, a liquid freezes all at once, a solution "precipitates"), fusion, sublimation. These reproducible and yet mysterious changes of state have inspired metaphors in all fields of knowledge: living things and the social order crystallize within chaos; resistances and compartmentalizations melt down in the heat of the moment; ideas and decisions precipitate abruptly. In "phase transitions," nature seems to clearly affirm itself as a power of transformation, capable not only of sliding toward disorder and indifference but also of making order and difference suddenly appear.

And certainly, following kinetic descriptions, our "good-sense" ideas about large-numbered systems will have to be modified. Gilles Deleuze reminds us in *Difference and Repetition* that, according to Hegel, good sense is a partial truth associated with the feeling of the absolute. We have to accept as only a partial truth the idea that chaos is inevitably subjected to the law of indifference and statistical compensation. Chaos can also, at what are called "second order phase transitions," become actual illegality, a chaos of fluctuations that no longer fluctuate around an average, because none can be any longer defined, but rather reverberate throughout the whole system, confusing that which the distinctions between macroscopic and microscopic had differentiated. This chaos evokes for us, as perhaps it did for the ancients, the unimaginable state that often precedes the establishment of order in traditional cosmogonic accounts. It is also the "chaos-cloud" that Michel Serres proposed to us with respect to Lucretian physics; chaos, stormy combat, a creative *turba* within which the clinamen can give birth to the stable whirlwind of things.

In fact, at the critical point of phase transition a gas is no longer, strictly speaking, a gas, but neither is it a liquid; droplets of water of all sizes develop, they can go from a few molecules up to a macroscopic number, on the same scale as the system, and they are intimately mixed with gas bubbles, also of all sizes; the fluctuations of density that express the formation of the droplets can take on macroscopic dimensions, reverberating their effects throughout the system; the correlation length—that is, the scope of the repercussions of a local event—thus tends toward infinity, all parts of the gas now being in contact, mutually "sensing" each other. The system thus reacts as a whole to what is happening in each of its regions. More precisely, the critical point corresponds to the state for which, whatever the scale on which one describes the system, whatever the threshold of dimension from which a fluctuation will be taken into consideration, the result remains the same: the coupling between the separated points of the system has the same intensity as the coupling between its neighboring points. Thus collapses the hypothesis at the basis of the very concept of a macroscopic state: the distinction between local events and global description.

     I have spent some time on the kinetics of matter to remind us of an element of reflection that we sometimes neglect a little too easily. We often speak of "dissipative structures" created by the amplification of a fluctuation to macroscopic dimensions. We are then forgetting the paradox hidden by these words, the conceptual upheaval: a fluctuation was what was by definition insignificant and without consequence; the possibility of its amplification signals the end of the tranquil generalizations of physics. Remember, nevertheless, that this is not a reason to mix everything up; the fluctuation in itself *does not cause anything*. Fluctuations are incessantly and inevitably produced in systems peopled by billions and billions of molecules with stochastic behavior. What counts is the specifically kinetic phenomenon of its amplification, the opportunity that this amplification reveals, and which gives way to an intrinsically collective phenomenon.

     One can characterize equilibrium phase transitions by a strong contrast between the transition itself and the state constituted by its outcome. If the process of crystallization is a process that can be assimilated to a supermolecular dissipative regime, since the system behaves like a whole, where long-reach correlations appear and intense irreversible activity is produced, its product, the crystal, is defined on the molecular scale and devoid of activity. The irreversible time of crystallization, creator of structures, has nothing in common with the unchanging eternity of the crystal. The crystalline order is a completed process.

     It is in this respect that, far from equilibrium, the situation can become quite different, since the transition will no longer necessarily result in a state dominated by Boltzmann's order principle, but in what is called a "dissipative structure." Far from being a transitory process during which, momentarily, the system acquires a collective activity, the transition undergone by far-from-equilibrium systems is only the first instant, the appearance of that which will stabilize as a supermolecular dissipative regime.

     Thus dissipative structures seem to prolong indefinitely the fertile instant of the genesis of structures. Within a dissipative structure, such as the

"Bénard instability" in hydrodynamics, even though the molecules only interact by way of short-reach forces, they nevertheless adopt a strictly collective behavior. The "Bénard instability," formed by macroscopic convection flows, appears spontaneously within a layer of liquid heated from below, beyond a certain threshold of the temperature gradient. Billions and billions of molecules converge in a coherent whirlpool formation, rather than moving indifferently in every direction. If we wanted to calculate the configurations that correspond to the different possible macroscopic states in the situation of the layer of liquid heated from below, the appearance and maintenance of the "Bénard instability" would constitute an event of quasi-miraculous improbability. This is also the case with the "clocks" that appear in certain types of chemical systems when the flow of reagents that feed them is sufficiently intense: the variation of concentrations in the system's reagents, which oscillate with a clearly determined frequency of macroscopic scale, constitutes a coherent, collective behavior, which involves a clear collective effort of "communication" between molecules that are essentially independent. Here again, the idea of a priori, equiprobable molecular configurations must give way to a logic of processes.

Up to this point, I have wanted to share the significance I give to the passage from a physics of states to a physics of processes, from a physics that deals with states of affairs to a physics that tries to recount histories. I am not going to spend time telling you about the variety of dissipative structures, the different diagrams of bifurcation, or of properties such as structuration from external fluctuations or the influence of external fields, nor of the possible connections with the problems of biology. On the one hand, these problems have been dealt with at some length in *Order out of Chaos*, and, on the other hand, to take up the distinction introduced earlier, it is a matter of purely theoretical problems, in which the specialists of mathematical representations of physicochemical systems speak of systems defined by physicochemical processes. Here there's no problem of legitimacy. However, this is not at all the case when modelizations address themselves to other types of systems, and it is to this question that I will devote the end of this essay.

What lesson can modelizations learn from all this? For a domain in which the tools of kinetics can have a certain relevance, notably when one is dealing with large, partially connected populations, the main thing is perhaps the clear distinction to be established between intelligibility and generality, the highly limited character of the validity of all the general functions that populate, for example, economics, sociology, or psychology, and that are constructed on the model of functions of state. Here I am alluding in particular to the different functions of optimization. I have no concern about illegitimate extrapolation while using physics in order to criticize them: these kinds of functions have their direct source in classical physics and produce in their domain the same veiling of processes. Independently of the methods that have been appropriated by each domain, it is difficult to imagine how any of them could avoid discovering for themselves what physics has discovered: that average values are not valid a priori but only within the limits decided by the functioning of the system itself; and that, in order to understand this functioning, one has to carry out a detailed investigation of the coupled processes that constitute it.

Another inspiration could come from the lifting of the constraints that have been produced by alternatives that are too rigid, such as between arbitrariness and determinism. The fact that physics can now describe regimes of functioning that are both *determined* and *open* onto the external world, integrating in a singular global function the rules that govern the transformations in the system, the present circumstances, and the past history of the system, obviously proves nothing outside of physics. Nevertheless, it is possible that the awareness of a certain conceptual vacuum will actualize this and cease being filled with the paradoxical association of disparate concepts. It is important here to remember that we do not yet have at our disposal any theory of organization. And if one considers, in the work of Jacques Monod, for instance, the crucial weight of arguments and metaphors drawn from, on the one hand, both cybernetics and information theory, and on the other hand, classical physics, one cannot underestimate the importance of the recognition by physics that the description of an "organized situation" is a quite open problem.

In this connection, I would like to make an observation that bears on the respective presuppositions of the models inspired by kinetics and cybernetics. Kinetics can, on occasion, lead to a phenomenon of self-organization, but never to self-regulation. Indeed, in no case will it attribute to such and such a substance the property of regulation. Self-regulation involves the possibility of passing directly from the local role of molecules to the global significance of this role. When one hears it said that "there is threshold of concentration beyond which molecule x enables this or that synthesis," when one attributes the responsibility for a regulation to such and such a class of reagent, the argument refers to a technical or social understanding of organization: what is being described is a circuit of which each piece has a functional identity defined at the level of the whole system. Each molecular type can then be held directly responsible for a certain number of effects entering into the global organizational logic. The kinetic approach, on the contrary, distinguishes quite strictly between the *feedback* properties that characterize certain stages of transformations of which a system is the site and the global *functional* properties of the system, such as stability or eventually regulation. The only grammatical subject for these properties is the system itself. One can speak, for example, of the catalysis of a reaction by certain molecules but never of regulatory substances or of regulatory feedback. It is the system itself that will or will not cross a threshold, that will or will not be stabilized, and that, eventually, will organize itself.

That being said, we must also emphasize the huge difference between the representations of physicochemical systems that can be defined under experimental conditions and these representations when they inspire models about the concrete living and social reality. It is for laboratory systems that physics has discovered the diversity of the regimes of functioning that are produced when the system is open, functioning far from equilibrium, that is, when one gives processes the chance of producing a result that is not simply a trivial equilibrium. But concrete systems are open in a quite different sense. Contrary to chemical systems for which we are supposed to take into account all the possibilities of reaction, living and historical individuals, cells, termites, or humankind whose collective behavior we can envisage studying are characterized by an indefinite multiplicity of interactions. Thus, a choice is imposed and the model can have no other value or validity than that of this choice. It is particularly important to emphasize this, given the prestigious transmutation that formalized language tends to impose on the most trivial choices.

Thus, in relation to the ensemble of neo-Darwinian models in which the evolution of a population is studied within a context of limited resources, that is, models based on the logistic equation, one cannot say that such models prove or legitimate natural selection on the pretext that they give it a mathematical basis. These models presuppose natural selection and present its consequences in very simplified circumstances. However, this does not mean that these models are without interest; on the contrary, they often allow real experimentation on the hypotheses and concepts that guide concrete explorations.

As soon as a choice is to be made between the interactions that one takes into account and those that are neglected—given the responsibility of "fluctuations," that is, the uncontrolled variability of individual behaviors—an inevitable risk occurs. Nothing guarantees that such a choice, appropriate in certain circumstances, will remain so in others, that the problem posed—that is, the very definition of the system and not only its regime of functioning—will not be modified. There is here a serious responsibility for those who create modelizations. They are always in danger of *ratifying* the definition of a system as it is given in the circumstances where they find it. By selecting, in their description of a system, the interactions that have been stabilized and privileged by the historical, social, and political context, they not only take note of this context but also justify it, because their models can only negate or overshadow the possibility of other behaviors that do not respond to the dominant logic. The responsibility is all the greater in that the time-scales characterizing the evolutions become smaller. When we describe the behavior of termites, we know that we are simplifying, and that the termites are, to all intents and purposes, capable of many other interactions. But we also know that thousands of years of evolution have privileged and stabilized the behaviors we are dealing with. On the other hand, when it is a matter of human populations, we are no longer dealing with millennia, but with lifetimes, or learning periods, or passing fashions. Consequently, the choice of the model is clearly a political choice. It is up to those who create modelizations to confirm a closure or participate in the exploration of other possibilities. It is also up to them to resist the theoretical jubilation that is completely legitimate when it is a question of natural phenomena, because there is indeed a clear jubilation involved in the conclusion: I can reduce this complex collective phenomenon, the construction of a termitarium, for example, to that small set of interactions. But the theoretician's pleasure takes on a quite different meaning when it addresses situations involving individuals who, by a more or less violent constraint, have been "modelized," that is, restricted to a small number of interactions.

I said earlier that the transformation of physics could hopefully render more distinct the contradiction between research activity and the socialization of this activity. This potential contradiction is nowhere more evident than in the difficult distinction between the reductionist and the analytical mind. One of the most important distinctions of the physics of processes is that analysis is not in opposition to singularities, but rather complements them. It is the analysis of the details of processes, of couplings, of the interactions in a system that allows us to understand the rich variety of differentiated behaviors that this system is capable of. Analysis does not necessarily conclude that a system "is nothing but this" or is "nothing but that," but rather can produce "all that," and perhaps many other things as well. On the other hand, when we see certain concrete modelizations at work, we have to conclude that reductionism is capable of surviving and protecting its power in any theoretical framework. In fact, its real force is not one or another theory, but the formidable possibility of ignoring, scorning, manipulating, and dominating.

Powers of Invention

F   I   V   E

Black Boxes; or,

Is Psychoanalysis a Science?

FOR MANY people, the question mark in the question "Is psychoanalysis a science?" only concerns psychoanalysis. As for science, we are supposed to know what it is and what it can do, even when its knowledges diverge. Consequently, some people refuse to call psychoanalysis a science because it is not "objective," while others try to win back the title of science by invoking the example of quantum mechanics. Some, like Karl Popper, have invoked the criterion of "nonfalsifiability"; in opposition to this argument, others have used Thomas Kuhn's notion of "paradigm." I will not follow such approaches. I will not speak here "in the name of science," nor in the name of any one science in particular. It is possible — at least that is what, as coauthor of *Order out of Chaos*, I argued with respect to the physics of far-from-equilibrium systems — that the conceptual and practical mutations that transform a science can have an effect of liberation and invention on others. However, no science can serve as a model for another by claiming to be able to determine the risks of its knowledge, its particularities, and its questions. Correlatively, I will spare myself the ridicule of "judging" Freud's theses in the name of what we "know" today, for example of criticizing the use he made of the second law of thermodynamics, or of the inheritance of acquired characteristics.

What name can I then speak in? In the name of the exigency that runs throughout Freud's work: that of founding a science. It has seemed to me, notably in reading the articles collected in French under the title *La technique psy-chanalytique*,[1] that this exigency, far from indicating a "scientism" that would consti-tute an irremediably dated dimension of his work, as so many contemporary com-mentators argue, constitutes an essential key to the invention by way of which today's psychoanalysts recognize each other, that of the "analytic scene." Thus, the question mark in "Is psychoanalysis a science?" is, for me, a double one. What has happened to the science that Freud invented? But first of all, what is a science?

Let's start with a completely phenomenological description. It is clear that one of the important characteristics of the activities that are called "scien-tific" is that they lead human beings to work "together" in a totally different way from artistic, philosophical, or even technical activities. There is nothing easier than to cite the example of isolated philosophers, addressing themselves to those who perhaps, in the future, will be able to read them. One can think of many painters or musicians who were scorned and ignored during their lifetime only to be recognized as geniuses after their death. Of course, as far as technical invention is concerned, the situation is more complex, but the figure of the solitary inventor is not just a fiction but history piously records the name and memory of those who, in their un-known workshops, were the first to perfect techniques that are today industrialized. Strangely, by contrast, the status of scientific precursor is not an enviable one: he had an "idea," but did not know how to "prove" it and convince his contemporaries. There are so many possible ideas, so many theories, whether they be crazy or wise! Apart from the history of mathematics — singular in that a proof, even when incom-prehensible to its contemporaries (see Evariste Galois),[2] can be recognized later, and lead to a judgment against the contemporaries who were unable to understand it — the history of the sciences can be distinguished from the history of ideas in that it depends essentially on the judgments of those that it studies. The "misunderstood geniuses," with the exception of the few that are traditionally quoted, are hardly recognized. And the historian who unearths a possible precursor from the background noise of ideas would be thought of as scholarly and meticulous, a lover of histories of no importance.

The fact that scientists work "together" has been largely under-estimated, or, more precisely, defined as a *consequence* of the fact that science, by de-finition, supposedly generates *objective* statements — statements, therefore, that are capable, in principle, of producing agreement among all those involved in it. More-over, this is why epistemological analysis tends to present us with an isolated scien-tist, faced with what it defines as the crucial problem of all science: "what is a scien-tific statement?" Epistemological analysis takes on the job of formalizing, or, rather, correcting, the criteria whereby each scientist is supposed to solve this problem, that is, of generating a formal definition of what makes a science a science. The passage of the isolated scientist into the general community of scientists will then be solved in a manner that scientists define as *trivial*, as a simple addition of individuals who must, more or less, verify that each one of them has submitted to the general discipline.

Now, I hold that this "trivial" operation in fact constitutes the crucial point that will give us access to the singularity of scientific activity. If we take seriously the description of stories belonging as much to the history of the sciences as to contemporary practices, and particularly the controversies aroused by any new proposition, we are obliged to conclude that the criteria of scientificity or objectivity that should allow these controversies to be settled *did not preexist them*, but are on the contrary a major issue in discussions between scientists. And this situation has not been changed at all by philosophers of science and the criteria that they propose. There is a well-known anecdote about this. Niels Bohr delivered a lecture on the new quantum mechanics to some philosophers who were in complete agreement and considered Bohr's propositions to be perfectly legitimate. But Bohr said that he was disappointed and explained that anyone who was not shocked by the implications of quantum mechanics had not understood it. In other words, quantum theory responded perfectly to the criteria of scientificity of the (neopositivist) philosophers, whereas it imposed on physicists an intellectual upheaval that their own criteria had not foreseen and could not understand.

The consequence of my proposition is that it is pointless to search for a noncontextual, general definition of the difference between science and nonscience (which returns us to the uncertain status of "precursor" in science). Of course, one can proceed by way of the absurd and refer to a distinction that nobody would argue with today. Everyone, apart from a few interested parties, would refuse to confer on astrologers the title of men of science, but would give this title to specialists of quantum optics, biochemistry, or statistical mechanics (or, maybe it is a measure of the originality of René Thom's thought that even here there might not be agreement). On the other hand, if we try to formalize what allows us to recognize the difference between science and nonscience, we will have the greatest difficulty in excluding parapsychology because in fact the specialists in this field try in every way to respond to the criteria of scientificity that are currently on the market. In vain, however. Scientists have decided that parapsychology is not scientific, and this decision has been taken primarily by all those who would otherwise be situated very close to it, namely, psychologists, neurophysiologists, anthropologists, and so on.

Conversely, we would no doubt run the risk of excluding a science that today concentrates the efforts of the majority of leading physicists: the theory of strings and superstrings. What is more, we need to remind ourselves that the Newtonian forces were rejected at the beginning of the eighteenth century as being "occult" and "nonscientific" by Newton's Continental colleagues, and that, among Darwin's most determined adversaries, one could include some of the greatest philosopher-epistemologists of his era.

The fact that we cannot define what a science is does not mean that we are dealing with a false question, devoid of interest. *Quite to the contrary.* Every scientist, regardless of how uncreative he or she might be, is confronted with this question, as was Newton or Darwin, as well as many others at a lower level. Any new measuring device or method of description raises the question of knowing whether the measurement is "correct," whether it can maintain the significance that is given to it, whether the description is adequate, and so on. As in the case of "fractals," whatever the scale on which one examines scientific practices, the demarcation between science and nonscience is discussed, but if this demarcation is to be capable of mobilizing all those whose work depends, in one way or another, on the answer, it *only takes on meaning in the precise context in which it is posed*, which does not mean that "ahistorical" or epistemological arguments cannot intervene, or even "transhistoric" arguments that have bearing on the tradition of the science in question. But here it will just be a matter of arguments. The recognition of an innovation as scientific—that is also, depending on the case, the modification of the reading of the history of a science, the lessons that it appears to authorize, or the transformation of the notion of "fact" accepted by a science—are truly creations, which produce the criteria on the basis of which the accepted innovations will be described a posteriori as "obviously" scientific.

Correlatively, it can be seen that any discourse about science involves the one who engages in it: this discourse is virtually *part* of the scientific activity that it seeks to describe, in the sense that it can, if the case arises, intervene as an argument during a controversy. I will take that into account in this text: if my argument is to be used, why not take advantage of it?

Thus, scientific activity is, for me, an essentially *collective* activity, which indissociably produces its own norms and the statements, problems, or instruments that respond to them. We still need, however, to understand what it is that links scientists and allows them to work together. This problem was also resolved in a trivial way by those who define science through its objectivity. The "object," inasmuch as it imposes itself on everyone, constitutes the connector. But how do scientists as I have described them avoid dispersing, each producing their own definition, doing their own thing? Why do they accept the verdict of their colleagues if it does not give expression to any preexisting definition?

First of all, the problem needs to be made relative. The proliferation of disciplines is not simply due to a harmonious division of labor that is itself the product of a general consensus. It sometimes happens that groups split up because there was no agreement between them and each one has conquered the means of surviving independently of the others. It can, moreover, also happen that history "reconciles" the adversaries in the name of what they had no knowledge of. Thus Mendel belonged to a community composed mainly of agronomists who, rejecting Darwinian theory, sought the causes of biological evolution in the mechanisms of hybridization. The "rediscovery" of Mendel's laws, in 1900, marked the possibility *for the Darwinians* of "recuperating" Mendel's laws, that is, of also drowning in oblivion the claims of the practitioners of hybridization, from then on considered as "doing genetics without knowing it."

What "links" scientists? In languages of Latin origin, there is a term whose etymology is promising from this point of view. "Interest" actually derives from *interesse*, "to be situated between." To say that what links scientists is interest is, therefore, a bit of a tautology. But it is a question of a tautology with quite subversive resonances. Have we not continually heard science spoken of as the disinterested activity par excellence, and of the "disinterested consensus" of scientists? The pejorative sense that the term "interest" has developed is a clear indication of our culture and reveals a dislike of history, and, ultimately, of those who construct it. From Plato, the adversary of the Sophists, passing by way of Kant, who founded a priori what Hume had tried to describe as a historical creation, and, up to our present day, epistemologists seeking in an ahistorical definition of rationality the guarantee of the validity of what scientists agree on, it has been the same search for a point of view that transcends history and allows one to judge the interests of humankind, the same distrust toward those who do not claim access to a reality in the face of which interests should be silenced and subjected.

As for myself, I would go so far as to affirm that no scientific proposition describing scientific activity can, in any relevant sense, be called "true" *if it has not attracted "interest."* To interest someone does not necessarily mean to gratify someone's desire for power, money, or fame. Neither does it mean entering into preexisting interests. To interest someone in something means, first and above all, to act in such a way that this thing — apparatus, argument, or hypothesis in the case of scientists — can concern the person, intervene in his or her life, and eventually transform it. An interested scientist will ask the question: can I incorporate this "thing" into my research? Can I refer to the results of this type of measurement? Do I have to take account of them? Can I accept this argument and its possible consequences for my object? In other words, can I be situated by this proposition, can it place itself between my work and that of the one who proposes it? This is a serious question. The acceptance of a proposition is a risk that can, if the case arises, ruin years of work. This is why such a proposition will be put to the test by those who are interested by it as much as by those who have an interest in seeing it rejected.

A proposition that does not interest anyone is neither true nor false; it is literally part of the "noise" that accompanies scientific activities, a noise that may subsequently become a signal, but it is the person who succeeds in achieving this that will get the glory. If ever the proposition of Jacques Benveniste — from now on "interesting" to the extent that it is always interesting to refute a thesis published in a journal like *Nature* — should survive the trials that would eventually actualize this interest, the homeopaths will not get much credit for it.[3] Clearly, like all other doctors, they will have to learn the — at last — scientific practices that these trials will have defined. The status of precursor is not an enviable one in the sciences.

A proposition that has interested and has been accepted as linking the work of a number of people obviously is not true in the absolute sense. It is true relative to the methods of testing, *but also to the relationships of forces* that prevail at a given moment, or that are organized around it. The interests of scientists are not "pure" and it is this, if it were the object of my text, that should introduce the theme of the multiple relations between scientific, social, industrial, and other interests. On this subject, I will limit myself to pointing out that the most heterogeneous interests are, contrary to belief, always capable of associating, and this is without doubt one of the sources of their bad reputation. It does not matter that you are interested in my proposition for different reasons than mine; from the moment that you accept the conditions whereby it interests me, you interest me. I will also recall that no proposition can satisfy everyone because every interesting proposition redistributes the relations of signification, creates meaning but destroys it as well, and can lead to defining a trait, property, or problem that interested many as an appearance or as secondary. Thus, any interesting proposition establishes in itself a relationship of forces. That is why in *D'une science à l'autre, les concepts nomades*,[4] I defined a scientific concept as always having two faces, one turned toward the phenomena whose examination it organizes, the other toward the scientists that it judges and places in a hierarchy, depending on the type of interest that they hold toward these phenomena. In this way, thirty years ago, the concept of a genetic program defined bacteria as the royal road toward the understanding of living systems and disqualified the embryologists, who had previously been the leading researchers and were now convinced of exploring in an empirical way a domain that would, one day, be subjected to concepts coming from molecular biology.

Thus, a proposition that has been accepted is not necessarily the object of the consensus of a community that preexisted it. It creates this consensus, as well as the community that corresponds to it. It excludes from this consensus, and defines as marginal or constituting as a dissident community, those who—one sometimes realizes this later—might have had excellent reasons for not accepting it, for not accepting as conclusive, from the point of view of their own interests, the tests that this proposition has satisfied.

The sciences, as I have described them, have no other guarantee of truth than the risks taken by those who practice them when they accept, that is, rely on, hypotheses whose rejection could, eventually, destroy the meaning and value of their own research. They depend on the problematic situation that challenges every scientist, who needs others in the sense that these others furnish him with what could give meaning and scope to his research, who must guard against the criticisms of others who will put his claims to the test, who does not exist if he does not interest others, does not convince them that it is worthwhile going where he suggests they go, or avoid a path that he has succeeded in challenging.[5]

Nevertheless, the question arises of knowing why scientific activity, as opposed to other activities that bring people together (such as politics), gives scientists the means of coming to relative agreement, to the reconciliation of disparate interests. The singularity of scientific arguments is that they involve *third parties*. Whether they be human or nonhuman is not essential: what is essential is that it is *with respect to them* that scientists have discussions and that, if they can only intervene in the discussion as represented by a scientist, the arguments of the scientists themselves only have influence if they act as representatives for the third party. With this notion of third party, it is obviously the "phenomenon studied" that makes an appearance, but in the guise of a *problem*. For scientists, it is actually a matter of constituting phenomena as *actors* in the discussion, that is, not only of letting them speak, but of letting them speak in a way that all other scientists recognize as reliable. In a well-known expression, Kant affirmed that it is not the business of the scientist to learn from nature but to interrogate it, as a judge interrogates a witness. This is a relevant description, but it confuses, as normative philosophers often do, questions of fact, questions of product, and questions of principle. The real issue is actually the invention and production of these reliable witnesses. No evidence pre-exists scientific activity. All the phenomena that we know of are overloaded with multiple meanings, capable of authorizing an indefinite multiplicity of readings and interpretations, that is, of being utilized as evidence in the most diverse situations, and thus also of being disqualified as evidence. The whole question is thus, for the scientist, to produce a testimony that cannot be disqualified by being attributed to his or her own "subjectivity," to his biased reading, a testimony that others must accept, a testimony for which he or she will be recognized as a faithful representative and that will not betray him or her to the first colleague who comes along.

So scientists work, work passionately, and their work, like the concepts that are their instruments, is always two-faced: they work their "object," but think about their colleagues, about the way they might counter or reinterpret the evidence, invalidate it or demonstrate its "artifactual" character. A scientist is never a "subject," alone before his "object."[6]

Sometimes evidence is accepted, but the reading, the interpretation of its testimony continues: thus, in his indispensable *The Pasteurization of France*,[7] Bruno Latour described Pasteur's work to have microorganisms recognized as witnesses that explain epidemics, and to have himself recognized as their representative. But the ensuing coupled history of men of science and microorganisms continues today, and our bacteria are now quite distinct from their Pasteurian ancestors. But it can occur that a testimony is recognized as in some way definitive, and that it becomes, for example, integrated into a measuring device that scientists agree to use without questioning the basis, the "theories" that it presupposes, and this can happen even in the case of an industrial apparatus to which, as citizens, they might entrust their lives. In this case, following Bruno Latour, I would say that science has succeeded in constituting a *black box*.[8] A black box establishes a relation between what enters it and what leaves it such that no one has, *practically*, the means to contest it. Indeed, this relation has been integrated into so many research programs that have furnished results that are themselves accepted that no individual could hope to interest anyone in bringing it into question. In principle, the contention would always be possible. And sometimes it happens and succeeds: most physicists prior to Einstein would have without doubt accepted as a *black box* the Newtonian laws of motion. The opening of a black box is not an impossible event, but one that is *highly improbable*. When Jacques Benveniste claims that water has a "memory," he must be aware that *if he were right* whole libraries of books on the theory of liquid states and thousands of experimental devices would be invalidated, and he must therefore expect to have, from the beginning, a maximum number of barely interested adversaries who have decided to consider that the evidence he relies on is simply an artifact.

The prestige of a science is incontestably linked to the number of boxes that it has succeeded in closing, which is also to say to the solidity of the tradition that unites its members, to the number of "facts" they accept, not with the indifference of a linguist accepting, for example, that the earth turns, but that they accept actively in orienting their research, controlling their reasoning, giving meaning and stakes to their hypotheses, determining the risks, and therefore also the interest of what they are proposing. And this prestige is certainly legitimate. But it is here that one must take care and be suspicious of the convincing, seductive character of what I have been describing up to this point.

If one continued trustingly on the path that I have just opened, one would end up ratifying as normal the actual hierarchy of the sciences, the accepted distinction between those that succeed in closing black boxes and those that are called feeble or "soft" because none of their statements avoids contention, because they have not succeeded in inventing any reliable testimony for which they would be recognized as the authorized representatives. One would also end up ratifying as normal the particular historicity of the so-called hard sciences: giving as the goal, in both senses, of all scientific argumentation the recognition by all competent protagonists—or, more precisely, recognized as such in this respect—of the "objective" identification of the witness. The presupposition that *in truth* the third party has become the reliable witness for one party against all others is indeed embodied in the practice of the so-called hard sciences. They proceed by retrospective separation, at the end of a controversy, between what everyone accepts as *objective* evidence, and what will be read and interpreted as subjective derivation, in principle disqualifying from the beginning those who have, in fact, been defeated. They create the principle in the name of which the defeated had to be defeated. And the epistemologists ratify this procedure, introducing, for example, the theme of "epistemological rupture" in order to disqualify still further those whose propositions are no longer interesting.

As I have said, to speak about science involves taking a stand. And it is on this point that I will situate my commitment. I will argue—contrary to the assumptions of epistemologists who consider an objective statement as a right to which any rational scientist can lay claim—that the possibility for a science to attain the envied status of a "hard science" is of the order of an event, which happens but which is neither decreed nor merited. Thus, I deliberately put into question the sciences that have tried to merit this title by mutilating their object (behaviorist psychology) or by forgetting it (mathematical economics), or rather, I refer their history to another register, where the dominant interests will be of an academic, economic, and political order. The hard sciences, which serve as a model for the others, which nourish the dreams and ambitions of the "founding" candidates of science, are not strangers to this register, but they cannot be reduced to it. They follow, expressing and recalling an *event*, the discovery of a way of constituting a phenomenon as a reliable and unexpectedly articulate witness, the discovery of an access that they neither deserved nor had the right to expect. The discovery of such an access is in most cases a surprise even for those who will make themselves the representatives of the phenomenon: it suffices to remember Newton verifying for years, before daring to announce the hypothesis of a universal force that explained the observable celestial movements; the chemists from the beginning of the nineteenth century discovering the operational simplicity of stoechiometrical ratio; Jean Perrin announcing, "Atoms exist, Avogadro's number can be experimentally determined"; Watson and Crick faced with the double helix and the unexpected possibilities of understanding that it offered to them.

Retrospectively, all these unexpectedly simple accesses have shown themselves to be rather more sly and complicated. It remains the case that the sciences I have cited have not deliberately started with the simple, hoping thereafter to have the means of moving to the complicated, nor have they begun by simplifying. They have discovered an expected simplicity, an event that I do not hesitate to consider as a gift: unmerited but punctuating a history that it transforms irremediably.

In most cases, scientists and epistemologists have been in a great hurry to explain this history, to show that the access was deserved and legitimate, the consequence of an ultimately rational method or interrogation. They have made the method, which ensues from the event, responsible for it, and have, as a result, obscured what is essential: *no one has promised us anything*, and in particular, no one has promised us that, in all the fields of knowledge, the same type of event will be reproduced. They have said nothing about what the notion of method dissimulates: the fact that all measurements are not of equal merit, that they do not all create meaning, that not all methodical interrogation *commits* the one who carries it out, or makes him or her run risks that will allow him or her to interest others in it, to articulate and proliferate other risky interrogations. This is because it is only a measurement that involves meaning that runs risks for the one who effectuates it, and designates itself as the potential heir of the event in which risk and meaning are irreversibly engaged. The behavioral psychologist does not risk anything in accumulating facts about the rat trapped in its labyrinth, but the facts he or she accumulates do not interest many people, and do not generate any *problem* for them.

To what extent is the *current history of the sciences* determined in part by the model of the theoreticoexperimental sciences that were born from such events? Which is also to ask: to what extent does the prestige of the status of hard science paralyze (or lead into a forward flight, which is basically the same thing) the sciences that have not been able to invent their own ways of attaining this status?

Paralysis: I will not hesitate to speak in this way of hermeneutic "countermodels," based particularly on the difference between "understanding" and "explaining," and in doing so consolidating the two terms that they put in opposition. In what sense are there pure explanations? Why deny that the symbolic language of the physicist gives him a form of "understanding" that is, for example, silently attested to by the notion of "physical sense"? The notion of explanation offers the theoretical sciences a purified image, aseptic and naive, that is incapable of taking into account the passion demonstrated by the different styles of physical explanation. As for understanding, it obviously cares little for the always complex relation that our languages create between "understanding" and "explaining," in claiming to base itself on the "intersubjectivity" of relations between humans. Doesn't every word we use and depend on when we believe we have "understood" carry with it explanatory models, of which some can even become the object of mathematical modelizations (see the theses of Thom and his school)? Pure understanding seems to me as illusory as pure explanation.

Forward flight: in principle, the successive failures that some sciences encounter in creating an object capable of mutually arousing, articulating, and implying the interests of a community could be in themselves *interesting*. The failure is, potentially, as a consequence of its irreversibility, an apprenticeship: one could, but one can no longer, think that…But, very often, those who propose a new attempt, a new foundation, consider the failures as errors, ideological deviations, and so on; that is, the failures are related to the inadequacy of those who encountered them. One more effort and we will at last be "scientific" is an expression that can be heard in the history of some sciences—as if the possibility of finally working together in the manner of the "hard" sciences simply depended on the (good)will of men, as if the failures did not constitute an inheritance that, if it was recognized as bearing on that which scientists deal with, could be shared, just as much as the successes of other sciences.

These last reflections situate me. In relation to the sciences, I am not, as I have already stressed, a "neutral" analyst. I will define my involvement a bit more precisely before moving on to the second problem of my text, that of psychoanalysis. I consider that the so-called modern sciences, born nearly four centuries ago, constitute a singular adventure, profoundly original, enthralling—and which enthralls me—because they have taught us both about the world and about the men who run it. I would like to point out that never, when one of these sciences has taught us something that we had not suspected about the world, has it disappointed us. Never, from the discovery of Newton's force of attraction up to the unimaginable jungle of neurons that inhabits our cranial cavity has it impoverished our imagination, but it has continually stimulated it to explore new paths. However, this history presents me with a challenge: how to succeed in "working together" where the "event" does not occur, where phenomena continue (and seem able to continue) to speak in many voices; where they refuse to be reinvented as univocal witnesses, as objects in the Kantian sense, that is, in the sense that produces a relation of judgment between the subject of scientific knowledge [*le sujet savant*] and his object.

       Some sciences have adopted this approach. It is not by accident that the American creationists have been able to show, with the support of epistemological arguments, that Darwinian evolution did not constitute a science. In fact, Darwin did not hand down to us a simple access to living beings. On the contrary, he destroyed what we had thought so, such as the stability of living species, the harmonious finality of ecological relations. He bequeathed us a labyrinthine world, unstable, patched together, a world that biologists should explore rather than judge. He offered us not reasons but hypothetical narrative frameworks. Freud compared Copernicus and Darwin by way of the symbolic power of their statements. But the heirs of Copernicus, the astronomers, specialists of celestial mechanics, and the heirs of Darwin, searching for exotic snails or even, thanks to recent genetic maps, the differential rhythms of evolution, the different components of genetic inheritance, have known quite different destinies. And yet (on this subject see the fine studies by Stephen J. Gould),[9] the specialists of evolution continue to learn, to discuss the partial and local reliability of such evidence, which is no longer considered as proof but as a clue.

       Learning to work together without this togetherness being centered on the production of objects, on the closing of black boxes; learning without being devoted to a faith in the guaranteed repetition of the event that opens up an intelligible world or in a method that is supposed to guarantee the agreement of the interested parties; learning also to discern the pretensions of those who, in the name of the scientific method, proclaim that their field will, by right, reduce others to it; learning to laugh at and make others laugh at reductionist strategies, which would be a matter of simple bluff if they did not succeed, as is often the case, in impressing research institutes and other sponsors, and thus turning into brutal facts the judgments they permit themselves; learning the humor offered to us by reliable and yet multivocal evidence, and, correlatively, the humor of interests that do not attempt to hide behind an objectivity in the face of which everyone should bow down: the humor of risky interests that entail a proliferation of constraints and of the questions that these constraints create; learning to recount histories in which there are no defeated, to cherish truths that become entangled without denying each other—this is what scientists are already doing in many places. It is what I work to have recognized against the politics of sciences centered on the myth that inspired the event-centered origin of the modern sciences.

It is now time to return to the question "Is psychoanalysis a science?" and make the question mark weigh on the first of these terms. Here I will have to resist a certain number of facile responses. Actually, on a first encounter, what strikes the external observer who reads the arguments that contrast psychoanalysts with those who seek to engage them in debate, questioning their interpretations or interests, is that, in one way or another, psychoanalysis claims the privilege, unprecedented in any other field of scientific knowledge, of not needing to give an explanation.

Of course, arguments like this lay claim to the model of the "hard sciences." In this manner, when the child constructed by psychoanalysis is contrasted with the child or children constructed by the ethologists who, today, try to observe the early interactions of the infant with the person taking care of it, it may often seem that the psychoanalytic child would have nothing to learn from other children. The psychoanalytic child would form part of the autonomous body of psychoanalytic theory, and could not, any more than it, be brought into question by the "observable" child.[10] The psychoanalytic child would indeed be "unobservable," like a chemical element that has little in common with the simple or composite bodies whose properties we observe but that originates from the theoretical interpretation of the whole field of analytic chemistry.

At first sight, a remark like this might appear to be healthy epistemology—but, I take the risk of maintaining, at first sight only. In fact, it presupposes what is here questioned, namely, that not only can psychoanalysis lay claim to the title of science, but, what is more, to the status of those sciences that are called "hard." It presupposes that psychoanalysis has succeeded in establishing, as much with what it interrogates as with those who might be interested in the same field, a relationship of forces such that no one has, in the present situation, the means to question the "facts" or the "theories" that it has established.

As I have already emphasized, "hardness" is not proclaimed, it is gained, as much on the level of the politics of knowledges, that is, the established distribution of the right to speak, as on the level of the testing that allows those who have been granted this right to work together in a reliable way. And the twin strategy (aimed at those with whom and that with which scientists deal) that promotes a science as "hard" must demonstrate its actual success through the proliferation and multiplication of interests articulated around its object. In other words, the epistemological right that has been invoked has no meaning unless it expresses the positive, historical fact that it *is recognized* by all those who would have the actual means to dispute it. No one disputes the electron of quantum theory because all those who could do this have need of measuring devices that involve this electron. A hard science cannot by nature be isolated (unless, like behavioral psychology, it "mimes" hardness). It does not have to defend itself against "neighboring" sciences, or affirm its theoretical autonomy with respect to them, because it has gained the means of organizing its connection with these other sciences.

Moreover, at the heart of psychoanalysis there appears to function a very curious "black box": the analytic scene itself. Doesn't one hear it asserted that the only people qualified to speak about it are those who have not only had the experience of analysis, but what is more, a true experience: to have been "badly analyzed" is as good an argument for disqualification as not having been analyzed at all. Whereas the black boxes that are closed by the hard sciences constitute theoreticoexperimental devices that confer a univocal and operational sense on certain facts, the analytic scene appears to create those who will have the right to speak about it, and therefore operates in itself as the foundation of right.

Of course, all training actually transforms those who undergo it. Thomas Kuhn clearly showed that the physicist's competence was not limited to theoretical knowledge, but also related to a "knowing how to ask questions" of a practical order. In the same way, the biologist's eye has to be trained in order to be able to read cells. But these different "know-hows" are not supposed to establish a limit *in principle* to communication. The physicist is supposed to be able to explain his approach; the biologist can draw or schematize the cell to show the inexpert what he sees. The knowledge that appears to be conferred by the analytic scene seems to be of a different order: it traces a limit that is quasi-ineffable and all the more insurmountable, between those who know and those who, not knowing, will, by definition, not be able to understand or discuss anything. There is here a profoundly singular state of affairs, which I will neither judge nor condemn, since, on the one hand, I do not practice normative epistemology, and, on the other hand, I am not one of those whom analysts recognize as having the right to speak—but it is a state of affairs that I am free to try to understand.

The perplexity increases when the external observer becomes aware, in relation to the controversies that divide analysts—that is, those that the analytic scene authorizes to speak—of texts like that of Robert S. Wallerstein, at the time president of the International Psychoanalytic Association.[11] We learn there that "theoretical pluralism"—the fundamental divergences that separate "orthodox," Kleinian, Lacanian, Kohutian, and other analysts—does not challenge the unity of psychoanalysis. Each theory, which is also to say each manner of interpreting symptoms, would constitute an explanatory metaphor, which the patient learns to accept and which will be effective insofar as it enables an affective and cognitive contact to be established between patient and analyst. The essential factor that unites analysts would then be nothing other than the analytic scene itself, the "interactional techniques constructed around the dynamics of transference and countertransference." In the same ecumenical perspective, other analysts propose to consider that different types of theoretical interpretation might suit different types of patients, or different stages of the same analysis.

Obviously, one could argue in objection to my perplexity that the metaphorical role of the different psychoanalytic theories is deeply analogous to that which is played, according to my own description, by the facts and theories that are set forth in the other sciences. Is it not a question of "interesting" the patient, of creating with him a manner of working "together"? Nevertheless, the difference is just as striking as the analogy, and it is so, within the perspective that I have chosen in relation to science, independently of traditional epistemological problems such as the difference between explaining and understanding, or the observer's involvement in what is observed. Indeed, scientific interests are designed to proliferate and diversify, to create networks that resist this diversity, to never cease reinventing its articulation. But, if one accepts the depiction proposed by Wallerstein, not only would no specialist (even from a closely related science like anthropology or the ethology of infants) have any reason to be interested in analytic theories, but, ultimately, analysts (whether they make use of a theory or move, in a pragmatic way, from one theory to another depending on the situation) could consider as "normal," *that is, as without any particular interest*, the theoretical differences that exist between them. Obviously, any theory is an instrument, but in this case we would not have an instrument confronting each user with the problem of its utilization and of that to which it is addressed, but a bag of instruments that coexist indifferently and that everyone takes possession of depending on the particularities of his or her own training or on the circumstances.

Here is, then, briefly summed up, the problem presented by psychoanalysis to the external observer. It must be admitted that the temptation is strong to stop here, to conclude that psychoanalysis plays on other registers than the interests of science, to conclude, for example, that analysts have no need to take the risk of arousing the interest of critical and exigent interlocutors since they have, as professionals and as mediating and cultural actors, the means of assuring their own reproduction. The black box, would, then, be rapidly described, both as an instrument of reproduction for those that it qualifies, and as the production of their means of existence. Psychoanalysis would thus be a job, a profession, but not a science.

Nevertheless, I will attempt, at my risk and peril, another path, a path whose possibility I discovered following research and discussions that I have been engaged in for some time with Dr. Léon Chertok.[12] I will attempt to take seriously Freud's claim of having founded a science, in order to ask the question of knowing what type of science this could have been. The hypothesis that I am now going to quickly develop, since it is addressed to readers who, on the whole, are supposed to know the history of psychoanalysis better than myself, should not be confused—and I stress this—with a global interpretation of psychoanalysis. It concerns a hypothetical reading centered on the question of the means that Freud invented to create what every science needs, that is, reliable witnesses, capable of intervening as such in support of theories that invoke their testimony.

First, I propose to take seriously the name Freud gave to the science that he intended to establish—"psychoanalysis"—as well as his explicit references to *chemical analysis*. This is a somewhat unexpected approach. Who is interested these days in the humble science of chemistry? Most of those who today look for analogies with the hard sciences prefer to search for them in relativity or quantum mechanics, sciences that they can consider as subversive as psychoanalysis. Of course, one could mention that during Freud's era chemistry was the queen of the sciences. I would rather clarify the stakes of my proposition.

Contrary to quantum mechanics or relativity, the reference to analytic chemistry cannot have bearing on a manner of description, a theoretical content, a lesson concerning the limit of our knowledges or their objectivity, but on an *operational technique*. Lavoisier, its founder, is less renowned for his theories than for his operational definition of a "chemical fact," which allowed him to claim the ability to overturn the relationship of forces between an individual and a tradition, that is, *to make a tabula rasa* of tradition.

The key word to Lavoisier's approach is *control*. The experimental "scene" of the reaction system must be perfectly controlled; nothing can enter or leave it without having been identified: the well-known set of scales symbolizes this control, but it was not enough. It was also necessary to attribute an identity to the products that entered into reaction, that is, to be able to say that they were *pure* (relative to the operations in which they intervened). Whereas the chemistry of the eighteenth century dealt with semipurified — that is, partially uncontrollable — products, analytic chemistry will broaden in scope during the course of the nineteenth century, to the rhythm of the protocols it will produce, with these protocols guaranteeing the production of pure products, controllable actors in new reactions, which will be able to be, in their turn, codified into new protocols that can be used by laboratories, but also by industry. The converging interests of academic research and industrial production during the nineteenth century forged the modern figure of this chemistry, in reference to which Freud baptised "psychoanalysis," a chemistry that, as Berthelot said, "creates its object," that is, produces the reliable witnesses of its theories, the reagents capable of entering in a controlled way into the henceforth intelligible reactions.

Tabula rasa of tradition, control, and purification: I would like to make these three terms, already brought together by Lavoisier, the thread for my reading of Freud's "technical" texts, that is, those texts in which the strange black box is invented and closed and from which, since then, psychoanalysts as well as their theories have appeared.

As you know, Freud left Charcot's Paris armed with a conviction and a hypothesis. The conviction, concerning which he will recognize his debt to Charcot, is that hysterical patients are not "malingeresses" — or rather, malingerers, since Charcot showed that hysteria was not a feminine privilege — and that this was the case even if their sickness could not be explained by an anatomical or physiological lesion. Where Charcot restricted himself to speaking of "functional dynamic lesion," Freud, basing his argument on the strange relation between hysterical paralysis and words — it is not the leg in the anatomical sense that is paralyzed, but the leg as we name it — would elaborate an etiologic hypothesis. Now, this hypothesis is basically concerned with the possibility of acting, of transforming. Charcot had demonstrated that, under hypnosis, one could induce artificial paralyses of a hysterical type. Words have the power to create, so why should they not have the power to cure? When, subsequently, Freud situates psychoanalysis in relation to the hypnotic technique, he will write that "Remembering, as it was induced in hypnosis, could not but give the impression of an experiment carried out in a laboratory."[13] Hypnosis, for the therapist, was an instrument, acting on the memory of the patient and allowing the patient to relive and speak the truth of traumatic memories, thereby becoming free from their burden.

As everyone knows, Freud abandoned hypnosis and suggestion, whether direct — "per via di porre" — or put in the service of the cathartic technique. Obviously, it is not possible for me to describe here all the events — the reappearance of the same symptoms, or the displacement of symptoms, the discovery of the amorous feelings that some patients manifested toward him, the abandonment of the theory of seduction and, more generally, of the idea that the "memories" hypnosis gave rise to are by definition true — that led him to conclude that hypnosis was not a reliable instrument, that it did not make patients *reliable witnesses of their own disorder*. The question is to know if Freud's renunciation of the hope of making hypnotic therapy a technique that had the reliability of a laboratory technique signifies that the institution of psychoanalysis marks the end of the analogy between therapy, from then on analytic, and the laboratory.

I will argue that this is not at all the case and that it is without doubt here that Freud's genius lies, the founding invention of a technique that would not only transform the obstacle into a driving force but also enable a tabula rasa to be made of all the techniques (shamanic, thaumaturgical) that the therapist, whether he wants it or not, inherits — while at the same time interpreting and replacing them with an intelligible and, above all, codifiable technique, that is, one that is transmissible and thus, in principle, practicable by anyone, like any scientific technique.[14]

I am not going to attempt to describe the obstacle (transference, which exposes the illusion that the therapist-hypnotizer might have about placing himself outside the problem, of only being the agent of its solution without wanting to realize that the patient constitutes the therapist, uncontrollably, as an actor in the problem's repetition) or the driving force (transference again becoming the pivot around which will be organized both the recollection of memories and the analysis of resistances opposed to this recollection). I will limit myself to an extended quotation that affirms everything my hypothetical reading requires:

The main instrument, however, for curbing the patient's compulsion to repeat and for turning it into a motive for remembering lies in the handling of the transference. We render the compulsion harmless, and indeed useful, by giving it the right to assert itself in a definite field. We admit it into the transference as a playground in which it is allowed to expand in almost complete freedom and in which it is expected to display to us everything in the way of pathogenic instincts that is hidden in the patient's mind. Provided only that the patient shows compliance enough to respect the necessary conditions of the analysis, we regularly succeed in giving all the symptoms of the illness a new transference meaning and in replacing his ordinary neurosis by a "transference-neurosis" of which he can be cured by therapeutic work. The transference thus creates an intermediate region between illness and real life through which the transition from the one to the other is made. The new condition has taken over all the features of the illness; but it represents an artificial illness which is at every point accessible to our intervention. It is a piece of real experience, but one which has been made possible by especially favourable conditions, and it is of a provisional nature. From the repetitive reactions which are exhibited in the transference we are led along familiar paths to the awakening of the memories, which appear without difficulty, as it were, after the resistance has been overcome.[15]

For me, this text speaks for itself: it explains Freud's strategy. Just as the eighteenth-century chemist no longer deals with the materials that he will use in the natural world, no longer studies the unpurified primary materials that the artisan transformed, but "creates his object," the analyst institutes a state that has all the aspects of an "artificial illness," whose only arena is the "circumscribed domain" of the analytic scene. "Morbid symptoms," the primary material of the former technique, must themselves be transformed, finding themselves given the signification of transference. By reorganizing the patient's neurosis around the analyst, transference renders it intelligible, accessible, as Freud says, to the intervention of the analyst since the analyst is supposed to have remained "neutral," not responsible for the roles that are ascribed to him, and therefore able to decipher these roles and demonstrate their meaning to his patient.

Thus transference enables Freud to substitute for the ordinary illness (which clearly involves the analyst, but to the same extent as any other character in the real life of the patient) a laboratory illness that refers only to the pure framework of the analytic scene. Transference enables the substitution of the uncontrollable illness by an illness whose transformed symptoms convey, to the analyst's ear, reliable evidence about what they express. The analytic scene has in this way become the quite singular *laboratory* where the substitution of the ordinary, uncontrollable neurosis must be produced by the analyzable transference-neurosis. The production *and* the analysis of transference thus assemble in the same process what the chemist was usually able to separate, since he found in commerce or in other laboratories the purified and standardized reagents that he needed. The analyst must manage at once both the process of purification and that of explanation, which is itself conditioned by the former process.

With the management of transference, Freud has not abandoned suggestion. Quite to the contrary, he thinks he has succeeded in transforming it into a controllable instrument: if "suggestion is the influencing of a person by means of the transference phenomena which are possible in his case,"[16] the difference between ordinary suggestion and analytic suggestion is the calculable and controllable character of the latter:

> You will understand too, from the fact that suggestion can be traced back to transference, the capriciousness which struck us in hypnotic therapy, while analytic treatment remains calculable within its limits. In using hypnosis we are dependent on the state of the patient's capacity for transference without being able to influence it itself. The transference of a person who is to be hypnotized may be negative or, as most frequently, ambivalent, or he may have protected himself against his transference by adopting special attitudes; of that we learn nothing. In psychoanalysis we act upon the transference itself, resolve what opposes it, adjust the instrument with which we wish to make our impact. Thus it becomes possible for us to derive an entirely fresh advantage from the power of suggestion; we get it into our hands.[17]

Far from being eliminated, suggestion is thus intensified, since everything that is opposed to transference is excluded, and its action is focused solely on the analyst. It is suggestion that allows for the closure of the analytic scene, its purification, the progressive transformation of everything that, in the experience of the patient, relates to real life into questions and symptoms centered around the analyst. Suggestion is therefore the *condition* for the convergence that Freud thinks he has brought about between the ambitions of research and therapeutic ambitions. Contrary to what is the case in traditional therapies "of transference," suggestion is not, however, therapeutic as such. It is the condition, not the reason: the cure, according to Freud, is not related in any way to the power of suggestion. The analyst's interpretations are addressed to the patient's "ego," and it is the patient that must be convinced that his symptoms of transference do not involve the real person of the analyst; it is the patient who must become aware of the resistances that his symptoms clarify. But it is suggestion that *conditions* the work on the resistances on the basis of which the patient will become a reliable witness, at once decipherable by the analyst and capable of accepting the implications of the reading that he has himself given rise to, without being able to escape into the indefinite and uncontrollable pretexts that real life offers.

Retrospectively, it is obvious to any analyst that the texts I have quoted refer to a blissful utopia. Nevertheless, the whole question is to know how this utopia will be characterized, which is also to ask how one will define the practice of analysts who can no longer adhere to it. Here we are on the dividing line separating two sides of the history of psychoanalysis, that of its foundation and that of its prolongation. Two divergent descriptions of the contemporary situation can be given according to whether the point of view of the foundation or that of the prolongation is adopted.

In particular, it appeared to me that two profoundly divergent readings could be given of the article "Analysis Terminable and Interminable," in which, two years before his death, Freud himself recognized the intrinsic limits of the instrument he had developed.[18] There is no point in summarizing this text, except to recall that here Freud draws up the list of reasons for the ineffectiveness of the analytic technique, the importance of the "battalions" that oppose themselves to the one who, thanks to transference, the analyst can mobilize. The transference-neurosis is not sufficient, Freud thus recognizes in 1937, to put neurosis "at the service of knowledge," to make it "accessible to the interventions" of the analyst.

From the point of view of the foundation, as I have tried to characterize it, this recognition has the character of a dramatic renunciation. No more than hypnosis or the other techniques utilized by Freud prior to the "foundation of psychoanalysis," do transference and its analysis have the hoped-for power necessary for constituting patients as reliable witnesses, as witnesses whose intelligible and calculable cure could confirm the validity of the theory that is supposed to confer its sense onto that which they suffer. Transference does not succeed in modifying in a decisive way the relationship of forces between the analyst and "real life."

Admittedly, Freud affirms that "qualitatively," the instrument is good: the problem is only quantitative. Nevertheless, it must be remarked that, in good logic, it is the instrument itself that distributes the categories of "qualitative" and "quantitative." Freud here situates himself on the side of "prolongation"; he does not reject the instrument as he rejected the previous techniques. On the contrary, he explains the relative ineffectiveness of the instrument by means of the theory, which has itself been constituted on the basis of this instrument. Most of the analyst readers of this text that I have been able to consult adopt the same point of view: they become attached to the elaboration of theoretical content that justifies the henceforth sometimes insurmountable character of the resistances whereby the patient opposes the analyst. The possibility of defining as a utopia the texts in which Freud defined the grounds for the analytic scene thus constitutes theoretical and practical *progress*.

Thus, Freud's text enables analysis to be qualified as an "impossible profession," in the best sense of the term, which is to think of therapeutic activity in terms of impassible limits. And this profession has discovered, since 1937 (you cannot stop progress), additional reasons for its "impossibility." In this way, the theme of countertransference, elaborated after Freud, has changed meaning: the well-meaning neutrality of the analyst is no longer an ideal that every analyst should try to attain (countertransference thus denoting a departure from the ideal), but a false ideal, countertransference becoming a legitimate inhabitant of the analytic scene.

The same difference of perspective, between the logic of foundation and the logic of prolongation, also shows itself with respect to the well-known theme of Freud's "scientism." If Freud's initial enthusiasm can be related to an outdated scientism, to the fact that he was born in the nineteenth century, the loss of this optimism marks the end of an illusion, that is, it signifies progress. An account like this can have two meanings. It can make the invention of the analytic scene the mark of Freud's genius (as one reads so often, Freud "knew how to listen" to his patients), and it thereby chooses to forget all of the justifications and reasons that Freud gave for this invention. The idea that a genius, almost by definition, turns out not to know what he or she is doing is a classic idea, a strategy that authorizes us to look down, from the heights of what we define as progress, on those whom we appear to venerate. But the concept of cunning reason is just as classic: it would have been necessary for Freud to claim to be scientific in order to invent the device capable (we know this now) of destroying the illusion of a possible knowledge, of a scientific type, concerning the unconscious.

It is not my business to criticize the idea that psychoanalysts have of their knowledge, and above all not to suggest that the analytic scene, stripped of the power that Freud had attributed to it, loses its meaning. That it is not the site that Freud thought he had constructed obviously does not mean that nothing special takes place there. However, the question "Is psychoanalysis a science?" leads me to consider as decisive the manner in which present-day analysts define the consequences of the "failure," declared in 1937, of the cure in the original sense that Freud had given to it. Indeed, it is actually on this original sense that depended his claim of having constructed not only a *singular* form of therapy, but also a scientific one, *that is, one capable as such of interpreting and therefore of making a tabula rasa of the whole of the therapeutic tradition that had preceded it* (a claim that, moreover, still accompanies it). In fact, Freud meant to establish a "hard" science, a science capable of creating its "object" and of judging it, and capable, through its operational difference from other practices that used the same type of instruments without understanding them, of judging those practices. If the thesis of scientism is adopted, that is, if the divorce between psychoanalysis and the exigencies of a "scientific technique" is considered as progress, can Freud's genius, or the quite traditional "cunning reason," preserve for psychoanalysis the privilege that Freud claimed for it? For me, the answer to that question is key to the response that is appropriate to give to the question about science.

In other words, there is nothing to say to the psychoanalyst who thinks he has a profession that is fascinating just as it is, but that does not enable him to judge other concurrent practices, and who, furthermore, admits that he has either no interest in these practices or one that is purely cultural. No one is enjoined to define oneself as scientific. I am addressing myself to those psychoanalysts who seek to uphold the claims of Freudian psychoanalysis, or who ponder over what type of science could ensue from the "failure" of 1937, just as psychoanalysis ensued from the "failure" of other techniques.

It seems to me that there are three possible types of response to the question of knowing how to defend the privilege that Freud claimed for psychoanalysis.

The first is of a cynical, pragmatic type. One can ask in what way it would be of interest for psychoanalysts to question the historical privilege that they have inherited. Unlike scientists, whose professional status depends on the networks in which they participate or that they contribute to establishing, which is to say the interest created by their work among their interlocutors, the analyst depends solely on his clients. He is what one calls, technically, self-employed. In this perspective, the idea that the analyst is privileged is not to be demonstrated, but rather defended, since the influx of clients depends, at least in part, on that. Is it surprising, then, that the divergent theoretical and practical propositions that actually coexist do not incite any crisis? Is it not better, as Wallerstein suggests, to spread the idea that the basis remains the same, even if no precise and generally agreed-upon definition can be given to this common property? To act otherwise would be to cut the branch on which all the analysts are sitting, weakening an image that they all support because they are all dependent on it.

The accusation of "cutting the branch," directed against some critical analysts, *has truly happened.* The cynical reading is thus not without relevance.

But its flaw is that, like any ratification of the status quo, it has no interesting consequences. There is nothing to learn from the history of psychoanalysis other than the eternal history of human credulity and the established conventions that humans invent in defense of their interests. Correlatively, interest here encounters its usual, pejorative sense.

The second type of response has been proposed by those psychoanalysts who have given up claiming that the cure is the application of a theory with explanatory pretensions. I am referring particularly to the interpretation of the cure given by certain advocates of what is called "narrativism," and also to those who confer on it the goal of pursuing a process of the "self-symbolization" of the subject. Here any claim to a truth that transcends the pair constituted by the analysand and the analyst can, ultimately, be abandoned (here I am not employing "transcends" in a metaphysical sense but in the practical sense of "being capable of interesting others"). The cure is a singular history that cannot, in itself, have theoretical bearing, because it does not have as a goal the discovery of a preexisting truth but the production of meaning, the invention by the subject of words and reference points that enable him to create meaning instead of being crushed by it. From this point of view, a theory would have none other than a practical function, the truth being related not to the constraints of critical examination, even through the cure, but to the constraints of *creation*.

However, the situation is not that simple. Couldn't one in fact say that any "cure," whatever the path to it, is at the same time a creation? On what grounds can analysis be distinguished from all the other techniques invented by human societies in the past and the present? Obviously, the statistics that tend to demonstrate that all the available therapeutic techniques can claim the same percentage of successes and failures are questionable to the extent that no agreement has ever been reached on the question of determining what constitutes a success and what constitutes a failure. But how is it possible, without a theoretical basis, to establish that one cure is better, or of a different type, than another? How can the paths of creation be compared?

In fact, it seems to me that the theme of truth as the creation of meaning, as an open process of self-symbolization, *is doubled*, in most of the articles that I have been able to consult, *by recourse to an other type of truth* that enables analysis to keep intact the claim of being able to judge other practices, that is, to provide a justification for the privilege of the analytic "cure." For me, this truth is of an ethicophilosophical order. It has all the characteristics of a philosophy of the subject authorizing an ethics that has bearing on the nature of the paths that allow this subject to respond to the human being's singular vocation. It is based, correlatively, on the denial of certain paths that one can clearly accuse the other techniques of using.

That being the case, analysis becomes philosophical anthropology, and not in the sense that Freud used in the latter part of his life, interpreting phylogenetic or even quasi-ontological determinations on the basis of psychic conflicts—such an interpretation presupposes the validity of analytic theory—but in the sense that the analytic scene itself constitutes the site where the truth of the human being has to be produced, in a quasi-structural manner. Thus, to take the example of the new "theory of generalized seduction" proposed by Laplanche as a *new foundation for psychoanalysis,*[19] at first there was the *enigma*, the unconscious sexual investment by the mother of her relations with her infant, who "feels" it and which becomes henceforth the source of this vague questioning: "what does she want from me, beyond feeding me, and, after all, why does she want to feed me?" What follows, as a consequence of the "enigmatic signifier," is a history in which, thanks to the logic of hindsight, unconscious seduction is the release mechanism for the *creation* of a subject *who is none other than that of psychoanalysis*, of a subject to whose truth corresponds trait for trait the singularity of the analytic scene.

The major characteristic of the transposition that moves us from the technicoscientific register to that of duty, from the definition of human suffering as posing the problem of its causes to the definition of the human being as having to be faithful to the (painful) problem that makes him or her human, is, for me, that it gives an ethical sense to the very thing that was, for Freud, a technical imperative.

As we have seen, Freud rejected hypnosis and did not modify the use of suggestion for ethical, but for technical, motives. It is difficult not to imagine that, if hypnosis, or indeed, direct suggestion, had been more powerful than the patients' resistances, the analysis of transference would not have been invented. On the other hand, from the moment that these resistances had shown their power, from the moment that they condemned to uncertainty any technique that short-circuited them, it became essential for Freud to show that the analytic cure was not the result of suggestion, that analytic interpretation did not work by way of suggestion but through its relation to a truth whose very sense was to incite resistances and prove itself through its capacity to overcome them. For psychoanalytic ethics, avoiding suggestion has now become an end in itself. The paths of suggestion could be used—they have been, and are used in all nonanalytic therapy. They *must not* be. Suggestion manipulates those who have to learn to speak for themselves; it drugs them with the words of an other, giving them the illusion of relief; it diverts them from the demanding path on which they have to learn not silence but to live the sense of that which provoked their complaint.

This second response is coherent, from a formal point of view. Nevertheless, it presents a problem that, contrary to Freud, it no longer has the means to resolve. Freud could, at the very least, define what was to be avoided since he was aiming at a truth that was defined, in some way, as preexisting, historical. We know that he undertook investigations aimed at confirming the validity of his hypotheses. But how is one to define this suggestion that has to be avoided? How — and what analysts have dared to say since Freud about the complex dynamics of transference and countertransference renders the problem even more insoluble — can one be assured that the analysand does not learn from the analyst what he or she is supposed *to create*? Also, how can one accommodate as analysts those who, like Kohut, appear to give a description of the cure where it is difficult not to recognize a deliberate use of suggestion (theoretically informed by the conception defended by Kohut of the genesis of the child)?

The history of philosophy and theology have already explored the paradoxes and reversals that affect the questions of good and evil, of grace and sin. Sometimes the most determined adversary of sin can be convinced of being the greatest of sinners; sometimes the one who seeks good with the greatest passion accomplishes evil through it; to believe oneself to be in a state of grace is a sin, and, in some coherent doctrines, to wish to be so is also a sin: is it not an insult to divine omnipotence? As for myself, the fact that the psychoanalysts (and particularly post-Lacanian analysts) can find themselves in the same type of situation with multiple and indefinite possibilities for reversal is the best sign of what analysis is here in the process of reinventing, and of defining as insurmountable: the question that, since Saint Augustine, traces a black path in Western history, the question of liberty as proceeding from an imperative of conversion faced with which all positivity, all immanence, and also all humor must be dissolved. Here, more than ever, we should reread Leibniz and Spinoza.

Who can guarantee that the intention of not suggesting is not the most unstoppable force of suggestion, against which the analyst has no protection? But, above all, what do we really know about this suggestion that we are supposed to avoid? Here we come to the third type of response that I will take the risk of proposing (or of suggesting).

I am speaking of risk here because I am now going to situate myself at the level that, as I have said, defines my involvement with respect to the question of the sciences. Its distinction from many other attempts to establish science that proceed by decree, by the unilateral decision of defining what the scientist deals with in terms that guarantee the possibility of theorizing and judging, is that, for me, the original Freudian attempt was not in any way "scientistic." On the contrary, it took into account, with remarkable clarity, what any hard science presupposes, that is, the experimental control of that which is interrogated. As such, it *could have* created the event, discovered an access whose simplicity in relation to reality was unexpected, in this case to psychic reality, an access that allowed this reality to be judged, that is, to construct for its subject a methodical approach defining an object. If—and we do not have the right to state that the hope was in vain—the description given by Freud of what the cure should be had been shown to be even approximately reliable, we would have progressively learned under what conditions the categories of the analytic scene can effectively *become* the principles of what, from then on, would have been an object in the Kantian sense of the term. As was the case with Lavoisier, Freud's science would have discovered the means of *creating its object*. This was not to be the case, and we now know what we did not know prior to Freud: transference does not suffice to make of psychic reality a theoretical object.

This was not to be, and, as far as I am concerned, I would not hesitate to see in the ethicophilosophical becoming of psychoanalysis a forward flight in search of new foundations that allow its claims to be maintained intact at the price of making them coincide with the question of the meaning of human existence. Of course, it is a rule of the analytic game that those who address themselves to an analyst are looking for the meaning of their existence. But this rule does not, for all that, qualify analysts to base the claims of their practice on the content that they give to this question, except, as we have seen, to close the circle and on this point disqualify all those who do not speak on the basis of the "experience of analysis."

The question, for me, is therefore to understand under what conditions the heirs of Freudian psychoanalysis could (like Freud, who at the time of the abandonment not only of the theory of seduction but of the conceptual and operational ensemble that was articulated around it, did not hesitate to do this) again take up the risks that are imposed on any science, although a trade or profession can, in a perfectly legitimate way, avoid them. It seems to me that this risk, which is also to say this new possibility of working together, occurs by way of a disjunction of principle between the two missions that Freud believed he had indissociably linked together through the power of transference: that of explaining and that of curing. This risk constitutes *putting into suspense*, even if only at a hypothetical level, all the discourses that describe Freud's creation of the analytic scene as an epistemological rupture and thereby ratify as unproblematic and irrevocable the judgments he passed on the instruments that he had renounced.[20]

It is not in the name of an abstract image of science that I am defining this obviously heavy price to pay (but, allow me to ask, Who more than analysts, given what they demand of their analysands, should be able to put into suspense the certitudes at the heart of their life?). And neither is it solely because this price is historically logical: it is in fact logical that the possibilities eliminated in the name of an apparently adequate and sufficient solution reappear when this solution has demonstrated its limits; it is logical, in particular, to ask oneself what hypnosis would be if it was rid of the illusion whereby the hypnotist is situated as an external observer of his patient; what is more, it is logical to again raise the question of knowing what suggestion can do in its many diverse modalities from the moment it is stripped of the illusion that the one who suggests knows what he is doing and can control the meaning and consequences of his suggestions with regard to the one he is addressing; and finally, it is logical that it is analysts, whose profession results from bringing to light these illusions, who are intellectually and affectively among the best qualified to take these new risks.

This last proposition makes clear the reasons for the hope that, as I have said, is my commitment. Psychoanalysts, as the heirs not of a science that mimes hardness, such as behavioral psychology, but of what *could have been* a hard science, are among those whom we could hope would invent new ways of working together that are not centered on the possibility of judging, but that enable us to learn how to learn. I consider that, as Freud's heirs, their responsibility is involved in having left questions like those of hypnosis or suggestion to the hands of behavioral psychologists, who, by definition, could not teach us anything. In a more general way, I consider that their responsibility is involved in not having been, up to now, the most critical and demanding allies of all the attempts that, since Freud, have readdressed the challenge of the "narcissistic wound" to the illusions of those who think they are in control, but, on the contrary, in having let this wound heal too easily; for they have maintained that, in exchange for this illusion, they had become the possessors of a knowledge, and particularly of a knowledge that allows for the judgment of those that have never shared this illusion.

For the narcissistic wound described by Freud does not affect the shaman, the mystic, or the thaumaturgist, but Occidental man, the white descendant of Descartes. Isn't it paradoxical, then, that, although the Freudian technique had disappointed the hopes of its founder, psychoanalysis confers on the white man the additional power that allows him to not only ignore them, but even to judge them theoretically? Isn't it strange that, precisely because, professionally, they have to be careful about the judgments of the white man, ethnologists distrust analysts more than anyone else because they are apparently capable of reducing a ritual trance to a hysterical crisis? Isn't it ridiculous that with respect to the phenomenon of hypnosis, whose enigmatic character Freud had always recognized, we are still at the level of invoking Hitler, drugs, or the music hall? Isn't it absurd that, with respect to suggestion, which has always been the symptom that we are perhaps not in control of ourselves, we have remained at the level of judgments whose Manichaean naïveté returns us to the era of witch-hunts? Isn't it striking that, in our era, doctors can still laugh at the placebo effect and analysts attempt to make a fantasy out of psychosomatic illnesses? Finally, isn't it worrying that psychoanalysts can pose the question of what could be the possible benefit to them of the attention and hypotheses of those who learn at their risk and peril, to observe the strangest human being that we know, the one with whom we can least identify ourselves: the child?

      The fields that I have just cited belong, it seems to me, among those that testify to the fact that the narcissistic wound of which Freud spoke has clearly healed too early, to the fact that new networks of interests need to be invented that propagate it and free it from its rhetorical role as the symbol and glorification of a profession. Doesn't psychoanalysis, being the heir to Freud, who knew how to confront the demands of a science and could have established one, have the vocation of taking the risk of participating in this invention?

Of Paradigms and Puzzles

THE QUESTION that I am going to discuss here might seem too general.[1] For me, it is, in fact, singular because it is inseparable from my own history. I do not know if I am addressing myself to you as a woman, as someone who, after studying science, could not accept the perspective of scientific research as she understood it, or as the philosopher that I have become. It is probable, although I did not think of it at the time, that the fact of being a woman played a decisive role in my decision to not engage in specialized research. But, to the extent that I have looked elsewhere for instruments in attempting to understand this choice, transforming a marginal situation into a practical activity, with its rigors and constraints, I can only speak here with many voices, without knowing *who*, today, speaks to you.

So, it is a question of identity—for me, but also, of course (how could it be otherwise in these conditions?), for that which I am going to attempt to talk to you about and which you know well, the sciences, and more precisely those that I am familiar with, the so-called exact sciences.

For some people, whom I will not waste my time talking about, the only identity that the sciences can have is that of reason. What may be already more interesting are those discourses that have attempted to define the singularity of science in terms of the desire for mastery, of manipulative, even violent, reason, of the a priori negation of anything that cannot be subjected to calculation and to the articulation between general law and a particular case. However, I am not going to dwell on this type of characterization, nor develop or comment on it. It is familiar to all of us. I will just emphasize something that has always seemed very strange to me, even though I used to accept it much more than I now do, and that is the very large number of essentially different authors who refer to it, and the essentially different vocations of the histories and analyses in which it figures.

A first name, Heidegger. For him, it is a question of the very framework of Western history, beyond anecdote, beyond histories, of what renders these anecdotes and histories intelligible. A second name, this time caught up in Herbert Marcuse's sphere of influence, himself a more or less deviant offshoot from the Frankfurt School: Brian Easlea. This time the sexed character of the scientific approach is emphasized, a date given to its origin: the rise of the bourgeoisie in the seventeenth century. In *Witch Hunting and the Birth of Modern Science*, Easlea shows that, for nature to be defined as an object of investigation, in terms that he shows are sometimes parallel to those that would describe a gang rape, it had to be stripped of any connection with the supernatural. Thus, it was also necessary to deny the very possibility that witches had supernatural powers. What interests Easlea is the actual transformation of sexed vocabulary in relation to science and nature: the feminine pole remains that of nature, but instead of being active and disturbing, it becomes passive and a completely penetrable object of control. He also analyzes the plausible social and cultural conditions for this transformation, with England and France being quite different in this respect. A third name: Bergson. This time it is no longer Western history or the bourgeoisie, its human intelligence as such, that privileges the foreseeable and the calculable. A fourth name: Meyerson. Here also, it is a question of understanding, but less of prediction and calculation than of identification. For Meyerson, understanding is essentially metaphysical. It is only satisfied if it has been able to show, in one way or another (and the history of these ways is the history of the triumphs of science), permanence beyond diverse appearances, equivalence governing change. In this case, the ideal of science is not that of a world that has been mastered, but, paradoxically, a world reduced to a cosmic tautology: Being or the Universe *is*. Finally, a recent experience that affected me: a trip to Japan, a meeting with Japanese philosophers and scientists at Tsukuba. And the discovery of their belief: that the paradoxes and difficulties of modern Western science express the dead end of the Western tradition, of the Western reason in relation to which Heidegger and Descartes are invoked, a reason that condemns nature to exploitation, the subject and object to separation.

What worries me from the outset, in this curious unanimity, is the splintered figure of the other of the identity of science as it is diagnosed. What is there in common between the Heideggerian meditation on Being and the social, sexual, and political liberation invoked by Easlea and Marcuse? What is there in common between Bergson's intuition, alone capable of approaching the creative singularity of things, and the Japanese tradition, which, as our hosts at Tsukuba told us, was going to take over from the Western tradition?

It is not by chance that I emphasize the chameleon figure of this *other* of the Western scientific tradition. We all know that among these *others*, there is also the figure of a science that would be the work of women, opposed to the competitive and calculating, dominant male. And yet again, what is there in common between the indictment of this Western tradition by certain feminists and its questioning by Japanese intellectuals, who, as one knows, are not celebrated for their feminism?

Of course, I am caricaturing, simplifying, schematizing. In doing so, I would like to express the deep uneasiness aroused in me by the binary characterization that the modern sciences seem to produce, constructing their identity through opposition with an *other*. And, in the chameleon character of this *other*, I tend to read the sign of the worrying amalgam that prevails in the reading of modern scientific rationality.

For me, the question is to try to break up this identity, to analyze this amalgam, and, in particular, because I think it is the most dangerous dimension of the problem, to dissociate the question of what one calls "scientific rationality" from the monolithic characterization that it has acquired as a result of its opposition to an "other." To dissociate does not mean to separate, but rather, in this case, to *suspend*. To suspend judgment, to recognize that what one presents as the object of this judgment does not respond to categories (is it useful to remember the strict relation between categories and indictment?) and therefore does not authorize any judgment. As a result, I would like to dissociate what is given as the object and conferred with an identity from the question of rationality, and thus, in the same way, to contest the possibility of speaking of an *other* rationality, whether it be intuitive, pre-Socratic, Japanese, feminine, or even proletarian (because, of course, the Stalinist opposition that used to be so much discussed in France between bourgeois science and proletarian science holds its untenable and dangerous character from the fact that "the passage to proletarian science" was presented as an ineffable quasi conversion to an *other* rationality).

This does not mean that scientific rationality can be called "neutral." Neutrality is still a quality. Without really controlling the implications of what I am putting forward, I would dare say that if we are going to speak of scientific rationality in a manner that is not superficial (as far as our problem is concerned), that is, that does not simply refer to the adequacy of a reasoning to recognized social norms, we would need to speak about it in terms of *jouissance*, with the abstraction, the absence of quality of *jouissance*.

In speaking of *jouissance*, I am not referring to this or that theory of psychoanalysis (that is also why I avoid speaking of "desire"). Obviously, I am not referring to a general characteristic, which would be common to all those who are called "scientists." There are "sad" scientists, and there are also "sad sciences," which mutilate what they handle just as much as the men and women who practice them. They can be easily recognized by the dominant reference to the notion of "method" that occurs in them. I am referring to a passion that one could call "abstract," because it is never encountered without being "qualified," without its being involved in a set of prescribed role games that usually confer on scientists their "serious," "objective," or "authoritarian" manner.

Obviously, there are scientists who avoid this role. Take the case of Barbara McClintock, the biologist who, in the eyes of some feminists, represents what a "women's science" could be: intuitive, holistic, respectful of what it studies. It is here that I need my "abstract" passion because, faced with this idea, I experience the same kind of concern as when confronted with the idea of a "proletarian science." In Barbara McClintock's work, one finds this passion that I am trying to characterize, which has nothing "reassuring" or "maternal" about it. McClintock tracked down the singularity of the genetic material of the corn she was studying, she defined it with precision and relentlessness. Her great *jouissance* was the moment when a "small detail" destroyed a grand idea, a superb generalization, when she knew that the corn had, if I can express it this way, "intervened" between her and her ideas. Of course, this *jouissance* has no connection with the unilateral, reductionistic, contemptuous spirit of domination, often and accurately associated with scientific rationality. And yet, she is associated with it on one point: she is essentially polemical. She does not commence with a pacified relation to the world, but with the search for ways through which the world can force us to abandon the ideas we have about it. McClintock was not interested in the social power conferred by the title of scientist. Thus she has never been "reductionist." But many reductionist scientists, stupid and arrogant outside their laboratories, may well seem like McClintock when one finds them in the laboratory, faced with an intriguing, surprising, or disastrous result.

It is not only the example of an exceptional woman that makes me want to take apart the amalgam that constitutes the scientist's *quality* and seek out its abstract characteristics, those which, like chemical atoms, do not exist in a free state, independently of the role assigned to them by signification and consistency, which makes of them either a defect or a quality. As for myself, I cannot ignore the *jouissance*, the occasional jubilation, of the researchers I have known. One can see in this a nostalgia for what I have denied myself. And, clearly, as I have said, I do not have the authority to identify *who* speaks when I am speaking to you. Perhaps it is the one who, in spite of the authoritarian and sterilizing forms of a scientific education, has actually known this intense jubilation of understanding how an equation works, or the way to trap a statement in natural language and give it the incisive precision of a mathematical statement, or yet again the tricks and detours, the bargaining over the nature of reagents, their conditions of use and order of appearance, which enables the synthesis of an organic molecule. Yes, without doubt, and any argumentation that seeks to establish the crucial character of this *jouissance*, this jubilation in scientific practice, could well be one way of giving this conviction, this experience, the attire of plausibility. And so what? The question is rather to know what such a thesis can lead to.

My thesis not only expresses a personal conviction, but it may interest others because of its effects, for it leads me to the following problem: the strange contrast between this hypothetical intensity and the manner in which scientific activity is presented. What, indeed, is rejected from contemporary scientific *literature*? The first thing that the apprentice author learns when writing a paper, a thesis, articles, applications for funding is: never say "I," but "we," never present research choices and methods as the expression of an individual choice, but as the expression of a unanimous and impersonal consensus. This is the reason for having lots of bibliographical references: you don't "do" your bibliography in order to initiate research, which in most cases is about to be completed at this point, but in order to show in what way this research is legitimate, in order to inscribe it in the logical and necessary path of scientific development. It is also unnecessary to mention one of the norms of writing that gives structure to a scientific article. Even in cases where the article's object has been a contingent product, and not the target of the research presented, this research is usually presented as having aimed, from the beginning, at the final result. And this result is (nearly always) used to redefine its problematic framework, in order to give it a sense that makes it rationally deducible from the reconstituted retrospective approach.

In short, it seems possible to argue that a large part of what is called "scientific rationality" answers above all to the norms of scientific communication. Rationality is not a category that enables one to speak of "science," because, if rationality is a normative category and not an abstract passion, it is thereby *a stake* in scientific practice that is continually being reworked and redefined. For a scientist, it is not only a matter of presenting his or her results as *rational* but also his or her choices. Barbara McClintock has been treated like a "mad old woman" not because any error or illogicality could be found in her work but because her methods and her object did not interest anyone. Indeed, she was not able, or more probably did not want, to have recourse to ploys that would make corn an obligatory reference point in the progress of biology. She disdained the continual double game of scientists. They will not talk much about interest, except informally among themselves; they will talk of premature reasoning, of wrongly based conclusions, of an experiment that was insufficiently controlled or interpreted in an ambiguous way. But they will hold up to scorn the efforts of epistemologists to formalize these criteria, because the decisive question is the propagation of an interest, of the conviction, whether or not it prevails, that here is a "worthwhile problem."

I have spoken of norms and representation. I do not wish to say that it is a matter here of external forms that leave scientific work intact. On the contrary, I think it is a question of determining factors that contribute to the production of science and to the way that scientific communities are reproduced. It is here that Thomas Kuhn's work interests me, not particularly because of the blindly dominating character of the *paradigm* that gives a scientific discipline its identity, nor because, in reading him, one sees that science functions in a dogmatic way, but precisely because a paradigm is not a dogma. Or, at most, it has the characteristics that a dogma would assume for perverse theologians who were only interested in it for the risk, for knowing what speculations are possible without infraction, and thus what redefinitions of the dogma in question are tolerable. Once again, it is a question of *jouissance*. Kuhn compares a scientist's *jouissance* with that of a puzzle solver. A paradigm is not a dogma, but belongs to science in that it only has sense through engendering puzzles, and thereby nourishing the *jouissance* of those who devote themselves to their solution (Kuhn speaks of "puzzle addicts").

So here is *jouissance*, but this time qualified—and qualified in a profoundly ambiguous way. Because what makes something a puzzle? Kuhn is specific: it is the conviction that the problem is soluble, and that its solution depends only on the scientist's skill. So far, the ambiguity has not been removed because this conviction can be given a number of different interpretations. Barbara McClintock's empathy, which enabled her to descend "into" the cells that she was examining, can come from this conviction. Yes, corn is intelligible, whatever may be the unexpected and fantastic nature of the interpretations that it forces on us. When McClintock was surprised by her corn, she laughed. A scientist, as described by Kuhn, laughs a lot less, and is not expecting any surprises, because, according to Kuhn, his conviction is based on his confidence in the adequacy of his method to his choice of object, that is, the possibility of explicitly extending the paradigmatic constraint to what he has "recognized" in the light of this paradigm, as a potentially soluble problem. Also, there are two senses to the word "puzzle" in English: one is "puzzled," intrigued, and open like McClintock, or else one confronts a "puzzle," a problem whose interest is conferred solely by the rules of the game.

I like pointing out this ambiguity because I think it is omnipresent. To be more precise, it is concealed and disguised by what I have called *norms* and *representations*. Everything is not for the best in the best of scientific worlds, because many of those who work in it do not deal effectively with this operation of dissimulation, do not understand it, or refuse to accept it. Who among us does not have in the back of his or her memory the blurred image of fellow students—I maintain that this does not specifically concern women—who, without having been especially persecuted, even in spite of advice and the general goodwill, even in spite of a recognized intelligence, did not manage to *integrate themselves*, to be recognized, to adopt the forms that enable this recognition, or the eccentric man or woman who was perhaps better than us even though we have survived them? I am not speaking of myself here because I did survive, perhaps because I was intrigued by what I have known. I am speaking of those men or women that one has lost contact with, who have disappeared from the field of visibility, after sometimes having been transformed into a caricature of themselves and having largely justified the exclusion that struck them.

The training of a young scientist is at bottom a subtle and ambiguous game, because it appears to contradict the very criteria that would finally allow us to distinguish the scientist who is renowned as a result of his or her originality and depth from the honest toiler, to distinguish them to the extent that the study of scientists' cross-references leads some to wonder to what extent science would be transformed if the grand majority of these toilers just disappeared: do the *real* scientists, those whose articles one reads and cites, this small percentage of the community, need their anonymous colleagues or not? Everything happens as if the "real" scientist was the one who deals successfully with being put to the test, who learns to conform and wins the right to decide how to conform. This right is not the right to be anticonformist but to become the author of an eventual innovation that can be celebrated as both the retroactively *obvious* thing to do and the testimony to the audacity and imagination of the one who did it.

I think that the discourse on scientific rationality is constructed in the continually reinvented link between "tradition" and "innovation." The innovative scientist must be able to be recognized as belonging to the tradition, as having accepted it, although his or her innovation will drastically alter it; for scientific work is not simply innovative, it is accompanied by a new account of the tradition that has been altered, an account that also establishes and renders plausible the innovation. Shortly before Einstein, the history of mechanics dealt with motion, force, and acceleration. After the special theory of relativity, new histories appeared, centered around the definition of invariants: Galilean "relativity," the symmetry of Maxwell's equations, and so on. Einstein's work could then be rationally deduced, and it was seriously asked *why* the unfortunate Poincaré had "missed" the invention of relativity. Surely he must have been the prisoner of a *false* ideal of rationality, both instrumental and conventional, not to have seen what had become obvious for all.

Can one imagine other regimes of functioning? According to Kuhn, it is impossible. In his way, Kuhn believes in the Hegelian "Ruse of Reason": scientific innovation needs norms and discipline; it produces its fruitful and shattering ideas because it creates a crisis in the consensual landscape that no one can ignore. From this point of view, the scientist, even if he has to destroy the consensus, must have been recognized as being a rightful member of the consensual community. His proposition may then be recognized and taken up by everyone, as if everyone could have produced it. Science needs order, and sporadic and innovative crises. A science that was in permanent turbulence, critical and open, would not be marked by any event. Perhaps we would learn more from it, but it would pose a formidable problem of management and evaluation. Imagine thousands of Barbara McClintocks! How would one rapidly gauge and evaluate the interest of a particular research if the interest of the researchers had not already been mobilized, aligned on some objectives that make them (nearly, and all the ambiguity lies in the *nearly*, in this "originality" whereby one recognizes the "real" researcher) interchangeable.

I am advancing a serious word in using the term "mobilization." It is not by chance that one speaks of scientific discipline. Discipline is the distinctive feature of an army that has been mobilized. Can we imagine a science that was not mobilized? Curiously, the closest memory that we can have of this goes back to

England during the Second World War. The memory of this experience is tinged with nostalgia for the many scientists who shared it because they were mobilized, not implicitly by professional consensus, but individually and collectively for tasks and problems that were not identified with competition. Thus, in order to decipher codes, not only were mathematicians brought together but also Egyptologists and crossword puzzle fanatics. The fact that war can liberate scientists gives one something to think about, but it cannot serve as a guiding ideal: it is an exceptional case, not a stable state of affairs. This is why the eighteenth century interests me. What happened to make the *Encyclopédie* so distant from us, or the dictionary of chemical terms in which Macquer attempted to present what he called the "facts," and which one could call constraints, leaving the reader-author free to trace his own path, free to evaluate the relevance of the theories proposed?

Some suggest that the *Encyclopédie* foreshadows capitalism in that it mobilizes know-how and resources, and attempts to give them a usable and communicable form. I think this ignores the fact that the mobilization of science by the state and by capitalist enterprise had quite different effects. The knowledges and know-hows of artisans were much less mobilized than invalidated; in this manner, nineteenth-century chemistry practically made a tabula rasa of the corpus that preceded it and dismissed it as still belonging to obscurantism (between the alchemists and Lavoisier there is "nothing"). This chemistry does not accumulate the same knowledge, and neither are those that do the accumulating the same. The science of the nineteenth century is constituted in opposition to the *Encyclopédie*, which argued for collaboration between theoreticians and workers; for the respect of the artisan's "flair," knack, and intuition. It is constituted by the invention of *enclosures*, where the ideal of the reproducibility of facts and the interchangeability of people can be approached.

It is interesting to reread Diderot's *The Interpretation of Nature* as an antidote, if one were needed, to the oversimplified identifications of science with "Western thought" or the "rise of the bourgeoisie." It is interesting because it inspires in us a healthy suspicion for the spectacle of unbroken continuity that sees the shortcomings of our science in Galileo's propaganda and in the "tabula rasa" operation that Lavoisier attempted on the chemistry of his time. It reminds us that we are above all heirs of the professional science of the nineteenth century. And perhaps it is on the basis of what began during the nineteenth century, and not on what one calls the "origins of modern science," that we should conceive of present-day science.

Let's return to the "identities" of science that we began with. They have a common characteristic, which distinguishes them from the "effect" that

Diderot sought in the *Encyclopédie* and discussed in *The Interpretation of Nature*. In both cases, it is a question of the "mapping" of phenomena into knowledge. But, with Diderot, the image of the labyrinth dominates: one can never move simply between two points that might have been thought to be near each other on the landscape of phenomena to explore; there are detours, but also long-distance analogical connections, dead-end streets, exploration occurs on the inside, without any a priori guarantee of relevance in the long term. This is why Diderot was, rather quickly, classified among the "empiricists." On the other hand, in contemplating the triumphant map that prompted the diagnoses that I cited at the beginning of my text, one might, on the contrary, think that one is in an already completely signposted country, covered with freeways and main roads. The distinguishing feature of freeways is to keep their relevance whatever the type of country being crossed: we know that they are not going to turn into muddy tracks or just come to a stop; we know that their network allows us to pass with ease through distinct regions and ignore the difference between plains and mountains, to pass, without any problems, from Brittany to the Mediterranean. So it is with "grand theories" that designate the real as intelligible through calculation and manipulation: they give us the impression that, even if we cannot penetrate the details, we can move from region to region.

Take the example of molecular biology. What does Jacques Monod tell us in *Chance and Necessity*? That there is a "freeway," a manner of crossing the landscape of the living without being, at every point, stopped by the singularity, the complexity of this or that living system. They can all be considered, on the model of bacteria, as the revelation of a "genetic program" formed by natural selection. Recently, we have realized that the "freeway" in question obviously allowed human beings to be integrated into the same landscape: why not go, without accident or solution of continuity, from bacteria to man, concluded the sociobiologists? And we have seen specialists of what they thought was a disconnected region, the human sciences, attempting to build barricades on the freeway, in order to prevent the influx of hordes of biologists transported without accident into what the specialists of the human sciences thought was their own property.

Kuhn's descriptions enable us to understand this manner of structuring the official scientific maps. What he calls a "paradigm," the generator of puzzles whose solution is guaranteed and which depends only on the player's intelligence, appears to me to coincide exactly with the claims of those who build freeways and highways. Use the intelligence of the network and you can approach any point already on the map; then you should be able to construct, one way or another, the remaining road ahead. One often hears of scientific "conceptions of the world." I think that it is less a question of conception, of vision, than of believing in road maps, the map of guaranteed accesses and their means of connection: in order to get there, you have to go this way. I think that this is in fact what a scientist says when he appears to deduce from a "conception of the world" that such and such a phenomenon is only this or that.

Mobilization also means mobility. The *Encyclopédie* also sought a certain mobility, but not that of an army in the field. It looked for interconnections, routes, the relations of expression between separated populations of phenomena, but not routes of penetration that string together at once all the phenomena and those who study them, while distributing a priori, regardless of the circumstances, what is significant and interesting, and what is only appearance or what can be ignored.

But are all the designated points unanimously accessible? The part played by tinkering about in the activity of scientists interests me. Although they claim to have *deduced* a singular description from the principles and general theories that enabled them to arrive close by, to have extended the network up to the phenomenon described, in fact, in most cases, they were only able to reach the phenomenon described at the expense of *approximations* that, sometimes, clandestinely contradict these theories and principles. Or, more precisely, they speak of approximations when it is a question of *tinkering about*, which is alone capable of creating the relevance of these principles and theories to the phenomenon in question, but which was neither authorized nor legitimated by the theory. We can pass through here, they conclude; the phenomena are unanimous here too, we can make them speak a common language. But this language has been clandestinely enriched by local constraints that will not appear in the official dictionary, and that will have to be learned *on the ground*. To continue with the geographical image, there are crevices on the routes to the different regions, and these crevices cannot be crossed without help. The local people have to lend a hand to help the traveler, and this aid is silenced in the official record of the voyage, whose primary characteristic is that of presumptuousness and lack of humor.

Finally, the problem of the contemporary sciences is not, for me, one of scientific rationality but of a very particular form of mobilization: it is a matter of succeeding in aligning interests, in disciplining them without destroying them. The goal is not an army of soldiers all marching in step in the same direction; there has to be an initiative, a sense of opportunity that belongs rather to the guerrilla. But the guerrilla has to imagine himself as belonging to a disciplined army, and relate the sense and possibility of his local initiatives to the commands of staff headquarters.

How can interests be aligned without destroying them? Here I would like to refer you to the works of Judith Schlanger, *Penser la bouche pleine* and *L'Invention intellectuelle*.[2] The question that concerns Judith Schlanger is the status of knowledge as interested. And in *Penser la bouche pleine* there is a theme that I find extremely powerful, that of the cultural memory, the multiple interests coexisting and densely interfering in this memory and in the natural language that we inhabit. For Schlanger, this memory, which goes beyond theoretical definition, nourishes our interest in that which we theorize and at the same time gives us the humor, the necessary distance to avoid being completely fascinated by the theoretical reading that reflects our presuppositions: it is this memory that "introduces the world between us and ourselves." And it is this memory that could enable us to understand what Kuhn says with respect to the scientist's initiation into the practices of his community: the lack of the history of a discipline, an essentially silent initiation, short-circuiting words in order to learn to recognize and reproduce model solutions. The role of cultural memory is thus actively and deliberately reduced and denied. It may, however, come back at times of crisis when scientists reencounter the use of words as they struggle to put into words what they believed they knew and discover the richness and ambiguity of language.

I would like to conclude with this idea: most of the diagnoses bearing on the identity of science have, it seems to me, taken for granted that this identity was deducible from something deeper, of which it was the irrepressible expression. On the contrary, I think that this identity, if it is able to characterize certain sciences (those that are called "hard"), expresses the mechanisms of extremely singular institutions. In this sense, it expresses nothing that would allow us to escape from our history and speak of science "in itself." And, consequently, it leaves us free to work at modifying these institutions without burdening ourselves with atemporal problems like those of Reason, Understanding, or the West.

S E V E N

Is There a Women's Science?[1]

BARBARA MCCLINTOCK is a scientist. Barbara McClintock belongs to a rare species of scientist; for nearly forty years she has carried out her research (which finally "had" to be awarded the Nobel Prize) in semireclusion, considered as an incomprehensible nut case by most of her colleagues. Barbara McClintock is a woman.

How should we arrange these three rough pieces of information? Should we, as is sometimes the tendency in feminist writings, give first place to Barbara, pioneer in the exploration of what might be a women's science, and therefore scorned and excluded by her male colleagues? Should we think of her story in terms of the rhythms of science, of "premature" works, misunderstood because they deviated from "normal" disciplinary research and subsequently recognized as precursors when this research imposes real problems that the consensus had until then defined as illusory or poorly formulated? Following her example, should we reflect on the tensions, the choices that set different styles of science apart from and against each other? Finally, should we see in the passion that inhabits her life a particularly intense manifestation of the specificity of the "life sciences," of those sciences that interrogate something that is not exactly an object, separated out from the confusion of things by human understanding, since it, likewise, organizes its interactions with its milieu and only exists by way of the multiple inventions of meaning and coherence that it inherits? As the great embryologist Albert Dalcq wrote:

In experimental biology, and particularly in the domains that touch on morphogenetic organization, deduction often requires a kind of art, in which sensitivity has perhaps a place. It is not only a question, as in physics, of acting on a variable, scrupulously recording the modification obtained, and recommencing until the numerical law is isolated. The very object on which the embryologist is working is capable of reacting, and the research readily takes on the appearance of a conversation: the riposte has all the unexpectedness and charm that one finds in the response of an intelligent interlocutor.[2]

The biography by Evelyn Fox Keller has the immense quality of not forcing us to choose between these diverse interpretations but of diverting us from any unilateral choice, teaching us to have the same respect for the concrete life of Barbara McClintock as she had for the organism, to distrust, like her, models on the basis of which one tries to explain *everything* while hanging on to what one thinks one knows.

Thus, there might be a great temptation to credit the solitary character of her fight against prejudice, during a period when feminism was dormant, to the fact that she affirmed herself less as a woman than as a singular human being, and as a scientist above all else. An ungracefully self-assertive human being, she protests when she sees her less-qualified male colleagues obtain "positions" when the doors remain closed to her, intensely aware that her difficulties come exclusively from the fact that she's a woman. Nevertheless, here it is the feeling of injustice that dominates rather than an awareness of her difference. Evelyn Fox Keller remarks that Barbara McClintock sought rather, to "go beyond the question of gender," and be recognized for her legitimate value as a scientist, much more capable than most of the men she had occasion to mix with.

Is it not possible, however, to recognize in the style of science that Barbara McClintock practiced, the difference that she herself denied? The way in which she describes her intuition to Evelyn Fox Keller, the empathy for the corn cells whose functioning she manages to *understand* in the most intimate sense of the term, within which she literally immerses herself—does this not constitute an exemplary illustration of the possibility of a nondominating, holistic science, capable of constituting an alternative to the reductive violence that some feminist discourses identify with "men's science"?

It is here that we must be suspicious of oversimplistic oppositions. Doesn't empathy signify a pacifying relation by definition? Isn't the most classic example of empathic rapport (perhaps that symbolized by totemic cults) the one that unites the hunter with the prey he is tracking? It is not by chance that I give this example. In *The Man without Qualities* (chapter 72), Robert Musil describes a very particular smile, which he calls "smiling into one's beard," the smile of the men of science who listen to the "celebrated men of the arts."[3] He writes that one should not see any irony in this smile; on the contrary, it is marked by homage and a feeling of humble incompetence when confronted by ideas that, quite clearly, transcend the scientific method. And yet, continues Musil, subconsciously this smile expresses what is brewing up in these men like a cauldron, "a certain propensity to Evil." Primal Evil that is finally "nothing less, nothing other, than the pleasure of tripping that sublimity up and watching it fall flat on its face." Who does not know, asks Musil, the malicious temptation—when contemplating a beautifully glazed vase— "that lies in the thought that one could smash it to smithereens with a single blow of one's stick"? But isn't this the same *jouissance* that shows through in statements that are called "particularly scientific," such as that which makes moral freedom an automatic appendage of free trade, or that which, to repeat Valéry, turns the poetic moon into a body that never stops falling toward earth in the same way as any old apple. Before intellectuals discovered this "delight in facts, the only people who had such a delight were warriors, hunters, and merchants, that is, people whose nature it was to be cunning and violent." Not being easily duped, bargaining without scruples for the smallest advantage, keeping a record of the minutest details in order to win the upper hand—these are, says Musil, the vices that science has transformed into virtues.

These vices-virtues of science also belong to Barbara McClintock and are what make her a determined scientist, tracking down the smallest clue, not accepting the slightest generality, however satisfying it might seem, respecting the corn less as a "totality" than as a confused labyrinth whose Arian thread needs to be found; not to "let the corn be," to parody Heidegger, the rather suspect reference of some feminist theoreticians, but to measure oneself against it like a subtle, complex partner, whose secret will only ever be uncovered by an effort that combines the minutest details and the imagination. And, of course, being a hunter, tradesman, or warrior are typically masculine activities, but I do not think that little girls are less apt than little boys to dream maliciously about the vase that everyone admires and that would be so easy to break. It seems to me that the real problem is the fact that these little girls, while growing up, learn to be ashamed and frightened of such ideas, while little boys are free to transform them into recognized and valued activities.

That said, it is undeniable that Barbara McClintock's science appears as singular, and not simply as "premature"; admittedly, it was put forward at "a bad time," at the moment when bacteria became the privileged object that finally enabled a bridge to be constructed between genetics and biological activity. But one cannot simply say that McClintock's only singularity was to maintain her interest for a multicellular while the consensus designated bacteria as the royal road toward the intelligibility of the living. In one way or another, it is also the way she interrogated this multicellular that condemned her work in the eyes of biologists, and the singularity of this type of science is in no way invalidated today, even though Mc-Clintock is recognized as the "precursor" of current descriptions of what the genome shows itself to be capable of. I would like to describe this singularity from two points of view, epistemological and social. Perhaps, as a result of this description, I will be able to reverse the proposition: Barbara McClintock did not practice a women's science, she was a woman doing science.

Two great theses have split epistemology since Hume and Kant. Does the ideal of rational knowledge designate a knowledge completely stripped of judgment, the collection of "pure facts" on the basis of which empirical regularities can be located and valid, logical generalizations constructed? Or rather, must it be recognized that there are no "scientific facts" without "man" who poses the questions, who always already interprets; that the facts, the moment they are distinguished, are therefore always "impregnated with theory"? Neutral facts or actively constructed facts. Man of science-tabula rasa, or judge, as Kant said, forcing the witnesses to reply to the questions he asks them. A third term escapes this never-ending alternative, which McClintock refers to when she says, regarding those researchers in molecular biology whose triumph condemned her to obscurity: "I'm beginning to suspect that a large part of the research has been done with the ulterior motive of imposing an answer on it. . . . If only we were content to let the material speak!"

Let's not be mistaken in thinking that this implies a return to empiricism; it concerns, rather, what the majority of epistemologists do not want to accept, what they agree, over and above their disagreements, to judge as irrational: the possibility that it is not man but the material that "asks the questions," that has a story to tell, which one has to learn to unravel.

One can see, in reading Evelyn Fox Keller, what is likely to give epistemologists the shivers. McClintock in vain describes her intellectual functioning in terms of a computer, integrating, comparing, and correlating multiple data: the very metaphor suggests a nonconscious, psychic activity. The *jouissance* itself is one of "not knowing" how one arrives, unexpectedly, at a hypothesis that one "knows" is accurate. Learning how to ask the right questions involves the dissolution of the conscious self, an opening that "lets the material come to us," but that signifies at the same time the abandonment of all the explicit intellectual procedures that enable epistemologists to construct models of rationality. They are, of course, used to distinguishing between the "context of the discovery" and that "of the justification," and Barbara McClintock also certainly does not content herself with subjective certitude but brings into play what she has learned and evaluates its actual pertinence. But the problem here is that the "discovery" cannot be assimilated to an initial phase, which could, a posteriori, be replaced by an approach conforming to an explicit methodology: it seems ineliminable. Faced with the confused jumble of "facts" that McClintock must confront, any logician would go crazy. It is only later, when the confusion has given place to a creation of meaning with the discovery of an Arian thread in the labyrinth of signs and clues, that one can begin to understand what Barbara McClintock "does," and also that she herself can explain what she is doing. Her very "explanation" also tells of the genesis of a conscious self out of a perplexity that involves indissociably the human mind and the corn.

One can find evidence of this type of science in every discipline. However, perhaps it more accurately designates what Fox Keller quite rightly calls a "naturalist" practice of science, a practice that does not proceed by way of a general judgment that distinguishes objects in a normative manner, defining a priori what they should be capable of or what type of question they should respond to, but addresses itself to a reality that is intrinsically endowed with meaning, that needs to be fathomed rather than reduced to the status of a particular illustration of a general truth. The example of Barbara McClintock enables us to affirm that this practice is not, as is often thought to be the case, antithetical to laboratory science. Unlike bacteria, the privileged objects of molecular biology because, at least on a first approximation, one could consider that they did not have a history, that their performances are simply the witness (that just has to be forced to speak) of their genetic identity, the corn studied by McClintock is the product of entangled histories, that of its reproduction, that of its development, that of its growth in the fields where it experiences the sun, the cold, predatory insects, and so on. Indeed, scientists should not accumulate "neutral" observations about corn, but learn from it which questions to ask it, because, like every historical being, corn is a singular being. And to say "corn" is already to say too much; for Barbara McClintock, each aberrant grain had to be understood in itself: not as representative "of" corn but more precisely in terms of the way it differed. Only later, certain general lessons would eventually be drawn, certain "principles of narration" could be defined that would enable an intelligible account to be given of all these singular histories, of these veritable biographies of grains of corn.

Evelyn Fox Keller enables us to evaluate the price of this "naturalist style": years of concentrated work. It is only possible to have two harvests of corn each year (as for bacteria, they reproduce in a few minutes), but this rhythm was still too fast for Barbara McClintock: so much evidence to gather, so many factors to relate together—one harvest was quite enough; and, of course, as a result, complex articles, judged unreadable by those who are used to the usual style of scientific publications (we have made this hypothesis, to test it we have devised this experiment, which produced these results, QED). These were detective articles: frenzied constructions, abounding in details, seemingly incoherent for those who only gave them a superficial reading; a slow and subtle creation of meaning, a narrative that could not be condensed from an encounter with a multiple and ambiguous adversary and that could only be understood by those rare few who knew that one does not "force" corn to speak, that one can only interrogate it by way of its distinctive and demanding conditions.

Epistemological theses, and the ideals of rationality of which they are the bearer, are only abstract in appearance. Their concrete relevance bears on what I will call the "social practice" of the sciences. The normative ideal, shared by the empiricists and the antiempiricists (from Kant to Kuhn), of a transparent construction of the operations of knowledge (based on facts, or on the relevant categories of human understanding, or on a disciplinary paradigm) implies, first and foremost, that any rational scientific proposition can be evaluated, judged in an economical way, become one piece among others in the edifice of knowledges. Barbara McClintock knew more about corn than anyone else in the world, and this knowledge, instead of opening up corn to an anonymous knowledge, making it accessible to researchers that one could ideally consider as interchangeable, accentuated its singularity: in order to follow McClintock's line of reasoning, one had to make the effort to become interested in corn, to immerse oneself in the multitude of problems presented by the smallest grain. This is why, with respect to her research, I have spoken of a "principle of narration," not of objective categories. The kind of intelligibility attained by McClintock does not allow one to forget about the concrete being, to reduce it to what it has allowed to be shown, but to recount its becoming, to understand, as with any real history, under what constraints each grain's history must have been possible, what was the influence of circumstances, what degrees of freedom they allow to be explored.

Concerning science, I have invoked the practices of the hunt. Now I must make a distinction. McClintock was a solitary hunter. The dominant epistemological theses are made for a hunting pack. The pack's principle is rapidity. The solitary hunter takes his time; he avoids any haste that the adversary-partner on the lookout would inevitably take advantage of. The art of the solitary hunt is empathy, a confrontation between the prey's subtlety and the hunter's. No one more than the hunter knows what his prey is capable of, can guess his possible ruses and foresee his initiatives and reactions. The practice of the pack is quite different. Here the prey is visible, panic-stricken, reduced to the channeled behavior imposed on it by the pack, whose members are ideally interchangeable. The main thing is the coordination between the behaviors of the participants, the fact that they all coherently understand the same signals. The pack *creates* an object accessible to "intersubjective" knowledge.

Nowadays, most scientists hunt in a pack. When one speaks of scientific reason as reductive, destructive of all singularity, and solely capable of reducing the singular to the particular case of a general rule, it is, I think, the practice of pack hunting much more than one or another identity of reason that is being referred to. When one says that analytical reason is incapable of understanding the "whole," forgetting that McClintock was a genius of analysis par excellence, one confuses reason with the irrationality that the practices of the pack are so often the bearer of.

Concerning ecological catastrophes, McClintock does not question "science" but stresses the fact that "we have not thought things through to the core of the problem." And if scientists and engineers do not think things through, it is certainly not because they are "rational," but because, from the point of view of the professional identity that defines their activity, *all problems do not have the same value*. Some are made to interest "anyone," to compel recognition because they are "worthwhile" for every colleague and for every sponsor. They constitute a genuine currency within the scientific community, whose circulation is essential to everyone, enabling the evaluation and organization into a hierarchy of those who pose the problems. In contrast, if you "think things through to the core of the problem," you do not allow things to be uprooted from the tissue of circumstances whereby they take on meaning, you do not allow things to be isolated, you do not allow judgments to be constructed from them that enable generalization, extrapolation, the oblivion of the thing for the comprehensible rule that can be used by anyone. In brief, you block the circulation.

Barbara McClintock's career was made possible by the fact that, initially, her questions interested "anyone." Corn, then, *helped* her to establish the general cytological identity of genes.

Competition, questions of priority: she was in advance of a colleague who accomplished *the same* demonstration using the drosophila fly. Barbara McClintock's singularity is that of a woman who succeeded in gaining scientific prestige and recognition during a period when, at the very best, women scientists could hope to teach in women's colleges. But, from the moment she chose to no longer make use of corn but to learn "with" it, Barbara McClintock made a choice that not only marks the history of women scientists but the history of the sciences themselves. There are few greater risks in the practice of science than that of marginality, of research that does not have the benefit of professional wardens, that is not controlled by the scrutiny of others but depends totally on the singular quality of the man or woman who undertakes it. We should listen to Barbara McClintock when she describes herself as a singular woman, beyond sexed genders. Her choice was that of a science in the singular, and this choice is as difficult for women as for men.

Nevertheless, Barbara McClintock was a woman, and that is not trivial. She had learned the art of solitude, the affirmation of singularity, the acceptance of marginality that literally makes so many scientists mad; she learned them in order to become a woman of science, to gain what would naturally have been given to her if she had been a man. Perhaps this is the real lesson of her life for those who are interested in the relations between women and scientific activity—not the discovery of an "other" reason, but the exploration of what reason is capable of when it is liberated from the disciplinary models that normalize it; the exploration of the real reasons one can have, even if one has a liking for it, for not feeling "at ease" in the sciences; the attempt, no longer isolated but interdependent and perhaps explicit, of resisting the social irrationality of the sciences.

E I G H T

The Thousand and One Sexes of Science

THE QUESTION that brings us together (do scientific theories have a sex?) is formidable.[1] Formidable, in the first instance, because of its global character. It concerns scientific theories in general, and not just certain theories—for example, those that deal, in one way or another, with sexual difference. Formidable, also, because it leads us to believe that we can define its subject: scientific theories. But do we know it? Can all scientific productions be considered as theories? How do we recognize a theory? Formidable, finally, because of the alternative it conceals. Either it is a question of asking if a scientific production can claim an *asexual* character, in which case, unless we accept the idea of *two* sexes, two distinct sexual *identities*, that would characterize conscious human elaborations in an ahistorical and asocial manner, a negative response to this question would allow us to envisage a thousand and one sexes for our theories. Or else it is a question of challenging the notion of "theoretical aim" as such, in which case we know what sex it would be a question of: it would be a question of the sex of power, which identifies knowledge with domination and ignores or destroys what it cannot submit to *objective* operational categories. But if, on the other hand, we have admitted that any scientific production can be considered as a theory, we then arrive at the formidable conclusion par excellence: no, science is not simply sexed like any conceptual production, it can be identified with *one* sex, with the one that affirms domination as a value.

This last conclusion is quite formidable, because it is akin to what I will call a *black hole attractor*, that is, a place toward which, if one does not take care, one is naturally and irresistibly drawn, and where one encounters, in the greatest confusion, other protagonists who have been led there by other routes. In this way, the feminist critique that would be led to assimilate science in general to a male perspective would find itself dangerously close to the cohort of Heideggerians who affirm that scientific rationality expresses the "historial" identity of the West, the metaphysical presupposition of a power as of right over being, of a world *made* to be represented and calculated. It might equally discover in its neighborhood the relativists, those who affirm that scientific productions have no other identity than that of the social power that distinguishes them and imposes them as true. It might even find in near proximity a few partisans of a "Japanese national science" who have also read Bacon, Descartes, and Heidegger and claim that it is from Japan that the world will receive a new type of science that will reconcile us with nature instead of opposing us to it.

How can we avoid being attracted toward this black hole while preserving the problems that appear to lead there? This is the question to which I will address myself here. And first of all, what is the common characteristic of all those men and women who have been drawn toward the black hole where science and power have been assimilated? I would say that all of them have sought an *identity* for science, which would have enabled them to *judge* it. Whether this identity refers to the male perspective, to Western rationality, or to social power is secondary. What concerns me is the common gesture — to identify — and the common position that this gesture authorizes: those who claim the power to identify, to separate the essential from the anecdotal, claim as a consequence the position and power of judges. It seems to me that the black hole brings together those men and women who have opposed the power that they denounced with what is then a superpower. In order to escape from the black hole, it is therefore essential to understand the powers at work in the sciences without making oneself the representative of a *superpower* capable of telling the hidden truth.

Privilege or Singularity?

"Science is different from all other practices!"

For many scientists, this is a heartfelt cry, a cry that needs to be heard, even if we remain free not to understand it exactly in the way that those who utter it would like. In fact, most of them, if asked to explain, would describe the "different from all other practices" in terms of privilege, and would distinguish science from other collective practices said to be stamped with subjectivity, instruments for the pursuit of different interests, guided by values that pose an obstacle to truth. Objectivity, neutrality, truth—all these terms, when used to characterize the singularity of the sciences, transform this singularity into a privilege. And this privilege, which confers on the sciences a position of judgment in relation to other collective practices, is also what the critics gathered together in the black hole transform, all in their own way, into an instrument of judgment against the sciences. Therefore, it will be a question here of attempting to think of the singularity of the sciences without transforming this singularity into the privileged expression of a rationality that would be set against illusion, ideology, and opinion.

Have we made any progress? The question remains general, but at least a trail has appeared. A direct consequence of the characterization of the sciences that we want to avoid is that it confers a particularly restricted sense on two traits that obviously contribute to their singularity: the sciences as a peculiar form of collective practice, and the sciences as defined by a peculiar form of history.

That the sciences are collective practices has truly passed into an unproblematic background. Objectivity, neutrality, the search for truth in a scientific sense are terms that condemn individual scientific activity and the collective product, the recognized scientific proposition, to be described in a homogeneous way. The collective character of the practice is "normal" because "real" scientists, each considered individually, agree to confine themselves to objective descriptions that can be compared with the facts, discussed, and recognized as adequate or not. In principle, "pure science," science relieved of the few *perturbations* that demonstrate that scientists are, after all, human, confers on collective activity the mere role of selection, testing, ratification, and transmission. In short, it identifies the collective instance as merely the collective recognition of "truly scientific" propositions— a rather limited role that is expressed, in their own way, by most epistemological doctrines when they question scientific propositions for the criteria that enable these propositions to be recognized as scientific, without examining the collective practices that incorporate these propositions, or when they take the possibility of these practices as nothing more than the attestation of the scientific character of the proposition, that is, of its fecund truth.

The fundamentally historical character of scientific practices has also passed into an unproblematic background. More precisely, the history of the sciences appears as if subjected to essentially ahistorical norms that will allow for the explanation of the criteria whereby the propositions are selected, tested, and ratified. The history of the sciences then becomes a history that, ideally, should be autonomous, a history that defines all other histories, whether social, cultural, or political, and with which it coexists, as a *context*, which, ideally, should be neutral, and which constitutes, depending on the case, a source of *perturbations*.

Both in the case of the collective and in that of the history, we find the same scenario: the pure, ideal case of a science that is in principle perfectly autonomous, of a collective reduced, on the one hand, to the role of recognition, and, on the other hand, to the perturbations, which indicate that scientists are, after all, only men, that there are, in the sciences as elsewhere, influences of fashion, power, and ideology. And in both cases, one finds the singularity of the sciences transformed into a privilege, affirmed by the necessity of beginning with the pure form and of understanding everything else as perturbations, as impurities.

Thus, for me, the question becomes: can we construct a description of the singularity of the sciences that avoids this type of scenario?

Fictions

I would certainly not have been able to construct the approach that I am proposing without the works of the French anthropologist of the sciences, Bruno Latour, but it will be impossible for me to define, at each step, what I owe to him, because here it is much more a question of appropriation than of borrowing. How can you define, in terms of debt, the fact of encountering the words and exigencies that, having come from another, enable you to progress further with your own problem?

This approach implies that one takes science seriously as a collective practice. This means, first and foremost, that one does not make the collective the simple outcome of practices that could, each individually, be judged as scientific or not (according to epistemological criteria, for example), but that one seeks to understand the collective process whereby an individual proposition is or is not *recognized*, that is, in fact, produced, as scientific. In this case, objectivity, truth, and neutrality would not be the qualities of an individual proposition that was the object of a collective recognition, but the qualities collectively attributed to an individual proposition when its recognition as scientific has been *produced*.

I am proposing to dare to call an innovative individual proposition a *fiction*. And I am doing this for two reasons: first of all, because it is clearly into the register of fiction that it will be rejected if it fails to have itself recognized as scientific; next, and more positively, because it seems to me that this term designates quite precisely a singularity of modern scientific activity, the liberty with which it treats what is given. To put it simply, whereas other traditions of knowledge have given themselves the rational task of justifying the given, of demonstrating that what is had to be, innovative scientific hypotheses always attempt to situate what is given within a much vaster set of possibilities. One can, in most cases, make them commence with "And if?" And if what seems to us to go without saying was not, in fact, as evident as it appeared? The scientific "And if?" is, by nature, corrosive. It attacks what we judge to be normal and commonsense. Or, more precisely, it expresses the fact that, at a given period, the judgment—this is normal—has become a little more shaky; it expresses and invents a positive meaning for the fact that it became possible, at a certain moment, to resituate an aspect of the familiar reality within a much vaster imaginary reality where what we know is only one story among others. A fiction, even if it is the product of an individual, always expresses what a history enables this individual to think, the risks that he is capable of taking. And here, we are truly dealing with the thousand and one sexes of the fictions that, at a given period, we are capable of.

To speak of fiction concerning an innovative scientific proposition does not mean saying "it's only a fiction." Propositions that contain the word "only" are all, by nature, reductionist. Those who voice this word are attributing to themselves the power of judging. The word "only" designates those who, in the controversy that an innovative proposition will arouse, have taken sides against it and made the wager that they will succeed in preventing this proposition from attaining a scientific status. And this expression appears in the common vocabulary in those cases in which such people have effectively triumphed. For us, the Lamarckian idea according to which the result of the adaptive efforts of living systems in an environment can be transmitted as such to their offspring is only a fiction.

Here are three examples that clarify the approach I am taking. At the origins of modern physics, this corrosive "And if" sounds loud and clear. And if what we observe always and everywhere, the fact that velocities naturally decrease, should be understood on the basis of a motion that does not exist anywhere in the world and that no one will ever observe, the eternal motion of a body that would go in a straight line, at uniform velocity? Nearer to us, Wegener dared to say: and if the obviously immobile continents were in fact like slowly drifting rafts? Nearer still, the biologist Lynn Margulis dared to say: and if the history of bacteria was going on within the history of multicellulars, and if we should understand ourselves on the basis of symbiotic populations of bacteria?

The three "And ifs?" I have just cited as examples now have different statuses.

Uniform motion has become so entrenched in our manner of thinking that some people ask why it was not always evident, and, scandalously, some school syllabi allow it to be presented to schoolchildren as directly derived from empirical experience. And yet Galileo's "And if?" was a scandal: to render intelligible what is given to observation on the basis of an intrinsically unobservable, fictional motion, which must from then on be recognized as that which allows the observable to be *judged*.

Wegener's "And if?" remained for a long time in the register of fiction. The evidence that, for Wegener, established continental drift, constituted for many others an argument that was far too weak—or, to be more precise, an interesting fiction. We "know" that Wegener "was right"; we know it because, since the 1960s, the movement (not of the continents but of the plates on which they rest) no longer has the status of a hypothesis bearing on the past, but has become contemporary for us: one can observe its effects, measure its velocity, reconstitute its chronology. We can criticize those who refused to "recognize" the truth of Wegener's proposition, but we must also measure the risk that his fiction represented for them: the disparate facts that become, as a consequence of an interpretation, the evidence of a hidden truth are the stuff of detective stories, but even in such fictions it is only the culprit's confession that can usually bring the case to a close (which is what happened in the 1960s when the plates confessed to their movement); in fact, we know (read, for example, *Foucault's Pendulum* by Umberto Eco) that understanding disconnected facts as evidence for one great interpretation is one of the places where genius and madness are the most intimately connected. Wegener was right, but many others remain in the memory of the sciences as "mad scientists."

The "And if?" of Margulis still hesitates today between fiction and scientific reality. But we already know the aesthetic, narrative, imaginative, and disciplinary upheaval that Margulis and those working in the same perspective as her want us to experience: with the long and slow history of bacterial populations, it is the soils, the oceans, and the atmosphere such as we know them that would lose their character of something given and become interdependent with the complex and hazardous history of Gaia. Yet another risk, because Margulis's fiction has quite clear ethical and political implications: as a scientist, can one accept a fiction that has the affirmed intention, among others, of engaging us in a history in which science and politics answer to each other?

Each of these three fictions has been met with the question, "Does he, or she, have the right?" or, in other words, "Is this scientific?" All innovative fiction in the history of the sciences gives rise to this question. Not "Is this true or false?" but "Can this support the question of true or false, in the scientific sense?" But what is meant by "in the scientific sense"? Here we are at the crossroads. Either we think that this question refers to an identity of the sciences—for example, to epistemological criteria that would determine the answer—and, in this case, the notion of fiction has not benefited us at all: we return to the notion of a history normed by epistemological criteria; or, and this is my proposition, the sciences never stop producing norms, but are not explained by them, because the meaning of these norms is one of the stakes of the transformation foreshadowed by the question "Is this scientific?" In this case, the question needs to be understood as the *problem* that is imposed on the collective by any innovative fiction, and through which the sciences invent their histories. To understand how the sciences respond to the question "Is this scientific?" is consequently to understand the singularity of an enterprise that succeeds in making human beings work together in a very particular way, that confers a collective sense on the corrosive dynamic of "And if."

A fiction that has the ambition of "being part of" science is therefore not just one fiction among others. It would not be enough to say that it has the vocation of being accepted by the collective; it would be better to say that its vocation is to transform the collective, and even, more accurately, *to create a new collective, which is to say, also and indissociably, a new type of history*. I am thus proposing to link fiction with scientific vocation, collective, and history, that is, to recognize that the wager of a fiction that has a scientific vocation is always to propose, even if only in a programmatic way, *a new mode for the intervention of a phenomenon in discussions between humans*, that is, to constitute a phenomenon as a witness, authorizing and supporting the thesis of the one who speaks in its name. In this case, a scientific fiction has the vocation of inscribing itself in a history and of transforming this history, of having the testimony of the phenomenon accepted in such a manner that it becomes a point of reference in this history, a constraint and the starting point for new fictions.

In this sense, the sciences can scandalize. Thus, for centuries, philosophers have assembled reasons a priori in order to demonstrate that the universe, inasmuch as it is exempt from the categories that define beings as affected by other beings, cannot as such be the object of positive knowledge, and, clearly, still less be dated in time. What would the post-1965 physicists, who have become specialists in the history of the universe, say to those who question them in this manner? They would say: "You're telling me that it's irrational to attribute an origin to the universe? But it's not I who am saying it; I only represent the residual blackbody radiation. It is confronted with this testimony, of which I am simply the interpreter, that you should admit defeat, you and all your a priori reasons. You will only defeat me if you can interpret this testimony differently and thus have yourself recognized as the legitimate representative of this radiation." That the universe has become an object of science signifies nothing more than the following historical and dated fact: the shift toward the red of galactic light and the residual radiation have been recognized, by a sufficient number of physicists, as legitimate testimonies of a history whose subject would be the universe.

Two questions now present themselves. First, if there is no ahistoric norm, how is it that the history of the sciences is not identified with an all too human history, where the strongest, the most convincing, the most seductive wins the day? Second, how is it that so many scientific theories, far from expressing the inventive dynamic that I have just described, seem to restrict themselves to mutilating what they are dealing with in order to subject it, in the name of science, to norms of objectivity? How can the formidable plausibility of the identification between scientific knowledge and domination be explained? Accepting this identification as a point of departure led us, as I have tried to show, toward a "black hole." However, knowing this does not mean that we can ignore the problem.

Histories

Let's start by attempting to respond to the first question. How can we avoid the history of the sciences being reduced to the arbitrariness of a pure relationship of forces?

Obviously, as is the case elsewhere, success in the sciences brings prestige, money, and praise, but perhaps a little less than elsewhere. This does not mean that scientists are "disinterested," as is often said. The history of the sciences is no more moral than any other history. This means that they are interested in something else as well. Clearly, scientists are rivals and not disinterested partners in the unanimous quest for truth, but it seems to me that their singularity consists less in living this rivalry in static terms (who, here, will be the strongest?) than in terms of history (who will have produced the testimony that will make history, that the others will not be able to invalidate, that they will all have to undergo?). In other words, what interests the innovative scientist is the history that his fiction will render possible and of which it will be a vector. And to the extent that this history implies interdependent works, which support and authorize each other, what interests him *is not the submission of others but the interest of others*. He needs others as innovators, creators of history, capable of taking new risks on the basis of what he proposes.

We find here, under a new light, the question of the *autonomy* of the sciences. The sciences are *autonomous* in that they delegate to no one else but interested scientists the responsibility for their norms, the definition of "what is scientific." But this does not mean that the process of definition is itself "purely scientific." It simply means that it is a question of the interest of scientists. And it goes without saying that this interest can, for example, be aroused by the marked interest of the state or an industry in the program proposed by the fiction, and that an innovative scientist can become a strategist, actively interesting multiple partners in his proposition. To interest is not the static consequence of a quality; to be interesting is first and foremost an active verb, implying the active intervention of new associations, new possibilities, new stakes. The passage from innovative fiction to a proposition recognized as scientific has interest as its key word.

It seems to me that the so-called *autonomy* of the sciences depends above all on the fact that interest is always combined with risk. Scientists do not have an answer to the ahistorical question "What is science?" but, on the other hand, for each and every one of them, the differentiation here and now between what will remain in the realm of fiction and what will be recognized as "scientific" is a vital problem. It is necessary that the associations continue, that the possibilities allow for work, that the stakes can be determined. As I have said, any innovative fiction is recognized by the fact that it makes a new phenomenon, or a phenomenon in a new mode, intervene in discussions. As such, it proposes the testimony of a phenomenon that will modify the degrees of freedom of all the works that must, or will be able in the future to, take this testimony into consideration, suppressing some and creating others. If a proposition has succeeded in being taken seriously, in being recognized as "scientific," that is to say, not as true, but as *susceptible of being true*, which is without doubt the most delicate stage of its history and the most fraught with arbitrariness, it is vital, as much for those who accept as for those who refuse it, that this proposition be subjected to trials that are all the more severe as the interest is keen, as the proposed transformation of the landscape of relations between things and man is important. The trial is the questioning of the testimony: isn't this an *artifact*, an extorted testimony that speaks of the device of interrogation, not of what is interrogated? Can't it be redirected, betray its representative, allow itself to be expressed in another language, authorizing other representations and therefore other representatives? These questions make up a large part of scientific controversies.

Scientific controversies do not have much in common with the usual ideal of rational debate. The protagonists do not seek an agreement on the basis of what they might have in common, which corresponds to the ideal of intersubjectivity; they do not attempt to base their proposition on evidence that the other, as a rational being, should accept, they literally demand, of the things that their proposition concerns, to witness in their favor. They do not address themselves directly to their adversaries but turn back toward things to try to invent the means of making them plead ever more decisively in their favor. In other words, they actively seek the means of making the history of their science not a purely human history, normed by human reasons, but a strange history in which things respond to man, a history that associates human arguments and the testimonies of things in continually renewed ways. "The order of things," one might say, with this difference that it is a question here, contrary to what was the case with Foucault, of actively created associations, of intrinsically costly associations, mobilizing laboratories, particle accelerators, observatories, studies, collections, expeditions, of associations that, sometimes, as Bruno Latour has shown with respect to Pasteur, involve a radical reinvention of both nature and society.

The sciences create associations, create, for better and for worse, new, formidable, interesting histories that continually stabilize the distribution of what is of the order of the actually possible and that of speculation, and they do it not only in laboratories, but also, if one gives them the opportunity, outside. But the question must always be asked: if the scientific becoming of a fiction seems to have for *effect* a transformation of society and of nature, that is, of the associations between humans, and between humans and nonhumans (things, animals, technical devices), who other than scientists had an interest in this transformation? Who gave the interested scientists the means to *make history* outside their laboratories, to create a world that from then on witnesses to the legitimacy of their propositions? Which are the propositions that *leave* the laboratories or offices of scientists, and which are those that remain confined there? The associations that are created between the sciences and politics are not secondary, sources of perturbations, they are intrinsic and of mutual interest.

Why is it that the history of the sciences can appear to so many of its critics as the monotonous repetition of a fundamental identification between knowing and dominating? Why, even though, as I have said, scientific fictions have in principle a thousand and one sexes, can one with so much plausibility assimilate the scientific vision to a *sexist* perspective? Until now, I have not responded to this question, but I have, I hope, shown in what way the singularity of the scientific enterprise is compatible with this state of affairs. Scientists are not, as such, "in the service of truth"; they cannot therefore be accused of betraying it; they are "in the service of history," their problem is history, and the truth, here, is what *makes history*. If the history of the sciences is dominated by a sad monotony, it is because, in our societies, those who can, if they are interested, give to scientists the means to make history are not just anyone, they are those who have the power to offer these means.

One will then say: the situation is hopeless. The sciences are doomed to be nothing other than a faithful expression of the dominant relationship of forces. That being the case, why this long detour, which appears to suggest that they could be other than they are? Because, it seems to me, this detour enables the critique to be transformed, to be made more *embarrassing*, and therefore more effective.

To move from the idea of a science that, by nature, assimilates knowing and dominating to a science that, in order to make history, exploits the multiple instances that, in our societies, are interested in the multiple inventions of such assimilations, is to move from essence to history, from Power to powers. It is thus also to no longer let oneself be impressed, to no longer be led to share the definitions of those with whom one enters into struggle, beginning here with the definition of science. Indeed, those whom one criticizes are highly unlikely to be embarrassed by the opposition between the sciences of today and *another* perspective, which would be, for example, that of women, because this opposition gives them what they are demanding: they are the legitimate representatives of "science."

Learning to laugh, in the name of the singularity of the sciences, in the name of the thousand and one sexes of their fictions, at those who give an identity to science, who say they know what the scientific method is, what the conditions of objectivity are, and what the criteria of scientificity are is a proposition that is in no way neutral. In relation to feminist movements, it presupposes, notably, accepting the idea that the women's struggle does not represent *another history*, but belongs, including in the creation of this reference to *another history*, to the skein of our history, in which the sciences are active ingredients. This, indeed, is my thesis.

Women do not simply belong to the skein of our history as excluded, minoritized, and slandered figures. They belong to it now more than ever because the question of women itself proceeds from the "And if?" of fiction. The question of women is political, cultural, ethical, social, and not scientific, but it belongs to the same family of histories as that of the sciences, these histories in which the truth does not mean ratifying a state of affairs, but subjecting it to the corrosive dynamic of what could be. This is why my position clearly does not amount to *defending the sciences*, but to defending their singularity in order to utilize it to invent the means of a critical position that *complicates* their history.

Theories

I will now give an example of what I consider to be a way of complicating this history in the name of the singularity from which it proceeds: the question of the right to theory.

From the outset, I posed the question: are all scientific productions assimilable to theories? Is any science entitled to produce theories? Who will answer "yes" to this question with the greatest certainty? The man (or woman) who will benefit from covering up the differences, and the risks that these differences imply; the man, or woman, who gives an identity to science, an identity that is largely inspired by a sterilized, *rationalized* version of physics, and demands that any other scientific production must conform to this model—even if this may mean mutilating, "in the name of science" or "in the name of necessity, for all science, of defining an object (of theory)," what one is dealing with as much as the men and women who ask the questions.

What have I been talking about up to now? Not about theories, but about propositions. I said that a fiction admitted as a scientific proposition has had a phenomenon accepted as a reliable witness, capable of intervening in discussions. Thus, to take an example, when Pasteur shows that for every contagious disease there corresponds the question of the transmission of a germ, he advances a proposition. But, on the other hand, when Pasteur, or the Pasteurian doctor, succeeds in having the germ, its identity, and mode of action recognized as the right question for anyone who is concerned with the scientific definition of the illness, *he proposes a theory*. Another example: when, during the 1950s and 1960s, the coded relations between DNA and proteins are recognized, when the code is unraveled, it is a question of a set of propositions. When one speaks of genetic information, and one defines the living by its program, it is a question of theory.

It quite frequently happens that proposition and theory, in the sense that I am in the process of defining them, are not explicitly distinguished. What I call a proposition, many would call a theory. The definition that I am proposing has the value of not relating the question of what a theory is to a question of epistemological status, but to the sciences as collective practices. According to my definition, one recognizes a theory by the *claims* of its representatives: they claim that, in such and such an outstanding case, the phenomenon is not restricted to testifying reliably, *but has testified to its truth*, which implies that to propose a theory is to claim to have conquered, in relation to what one is interrogating, a relationship of forces such that one has been able to make a phenomenon admit to its truth. At this point, the phenomenon is no longer only a witness, but becomes an *object*. To the notion of object corresponds that of *judge*: one can only speak of an object when one claims to know how to judge, to dispose of categories that allow one to distinguish between the essential and the anecdotal.

Together with theory, value judgments appear at the heart of science, and two distinct types of power are always entangled with these value judgments.

Power over the things that can be judged, such that one can now anticipate in what way they will have to testify. Thus, if bacteria testify to the genetic program, all living beings become, at least in principle, objects of the same theory: if I know how to judge them, what questions to address to them, I know in what way they should testify in order that their testimony expresses their truth.

Power too over the humans who are interested in things because one can now distinguish between the questions that are asked of these things. If a germ is the truth of a contagious illness, the fact that, exposed to the same germs, some people fall sick and others do not becomes secondary, and preventive medicine clearly becomes a useful, but not fundamental, research field. If the genetic program is the truth of the living, the essential is the resemblance between a bacteria, an elephant, and a man, all genetically programmed. What distinguishes them may certainly be interesting, but it will have to be redefined on the basis of the notion of genetic program. Embryology, the science concerned with the trait that differentiates the elephant from the bacteria since bacteria have no embryo, was the leading science during the first half of the twentieth century. With the rise of molecular biology, it became a set of unreliable empirical results, waiting for the moment when biologists would succeed in making embryological processes testify to their essential relation with genetic information.

All theory affirms a social power, a power of judging the value of human practices, and no theory imposes itself without, somewhere, social power having played a part; because, contrary to a proposition—which can drastically change and subvert the conceptual landscape, connect regions and disconnect others, but which defines the possibilities available to all, the constraints that everyone will have to take account of, but that everyone, if they invent the means, will be able to take advantage of—a theory requires that the hierarchization of the conceptual landscape that it proposes be socially ratified. Such a science, which poses the essential questions, is the leading science. Other sciences can be useful, because the questions they address to an object may prepare the ground for the leading science. On the other hand, these other sciences must be denounced as parasitic, ideological, or nonobjective because, if the questions that they pose and the testimonies that they seek were taken seriously, they would bring into question the theoretical object, implying that the resemblance affirmed by the theory is not essential, that some of the phenomena belonging to the field of the theory testify to another type of truth.

Starting with this essential relation between theories and power—power over things and power over those who are interested in these things—I do not want to draw a conclusion but take the risk of playing on the differences in a way that complicates the relation between scientific production and theory.

Scientific theories are not all the same. The notion of scientific theory, in the modern sense of the term, was created in the field of physics, and appeared there as if carried by the unforeseeable, practically scandalous event that the possibility of assembling apparently disparate phenomena within the same understanding constituted. The falling apple, the moon, the earth, and Halley's comet all bore the same testimony, the testimony of the same force of attraction! Since then, the history of some sciences has been marked in this way by unexpected events that we usually call "discoveries." Of course, no discovery is pure, and phenomena never spontaneously testify in the same language without efforts on the part of those who interrogate them. Physics is not only singular because of the events that have marked it, but also, and especially, because of the enormity of the pretensions and power that it authorized itself to lay claim to in their name. It remains the case—and this is why I speak of an event—that what has actually been obtained was far more than could be rationally expected, as if a type of question suddenly gave perhaps not a power over phenomena as radical as the corresponding theory claims, but a surprising power nevertheless. In order to characterize the theories created from such events, I would say that they are obviously linked with a social type of power, but that they are also born of the wonder of the event—the wonder of Newton, the wonder of Jean Perrin counting atoms, the wonder of Rutherford and Soddy linking radioactivity with the disintegration of atoms, the wonder of Watson and Crick confronted with their model of the double helix.

On the other hand, it has to be said that the very existence of some theories depends on social power. These are the theories that do not present themselves in the name of an event, but in the name of science, in the name of the principle of scientific rationality, constructing their object in a unilateral way, that mutilate what they are dealing with by eliminating a priori anything that does not appear to guarantee an objective approach, that demand of those that practice them that they mutilate their questions and interests by eliminating anything that will not respond to the conditions of scientificity.

Am I in the process of proposing a hierarchy of the sciences to you? No, in fact, I will go much further. I am in the process of trying to get you to demand explanations from the sciences that claim to have theories—not to ask them if they are independent of all social power (you will not find any pure ones, so the question is wrong), but to ask them what authorizes them to think that the reality they are dealing with can be judged, can become the object of theory. I maintain that those that reply more or less as follows: "Every science has its object," or, "It's a theory because it's an objective description, generalizing systematically observed facts," or "Obviously my theory is simple, but science must start with a set of simple and assured facts to be able to then make progress," are not, as one sometimes thinks, representatives of immature sciences but of practices that mime scientific activity. They have replaced the corrosion of the risky fiction with the unilateral decision to reduce what they are dealing with to the image of what they think is the "object of science."

Styles

The distinction that I have just introduced is itself a risk, a type of approach that engages the man or woman who practices it, which is to say that it is made to arouse controversy. What does this controversy aim at? To put into crisis the idea that the possession of a theory corresponds to a general right to which any field that merits the status of science is entitled, to demand from every scientific field that it pose the problem of the power it claims, where today reigns the idea of a right to power. Thus, strictly speaking, it is not a question of a scientific controversy, but much more of a political controversy, in the sense that it is directed at one of the judgments that today organizes the practice of the sciences, bears on the controversies, confers the appearance of an a priori legitimacy to certain propositions and denies it to others as being nonscientific or even *irrational*. But the stakes of this political controversy do not involve a general political claim but a question created by the very existence of the scientific enterprise in its singularity, the question of the interests and passions accepted as "being part of" "science."

Putting into crisis the idea that theory responds to a right held by science is, for me, not to propose a new model of science but to engage interests, questions, and passions that have until now been defined as outside of science, and that, in most cases, have been convinced to claim themselves to be so, to recognize the trap in which they have been confined. Obviously, nobody is forced to want to do science. But one thing seems clear to me: the question of knowing what the sciences can do, where they will be tomorrow, involves the question of knowing if new passions, new exigencies, and new questions will tomorrow accept the risk of "wanting to be part of" science.

For me, it is crucial to emphasize that invoking a political controversy in no way implies a utopia that envisages a desirable reality for which there is no evidence of possible actualization. *We already know* fields where new interests and passions make history, where a new style of *working together* has been invented. But we do not always recognize sufficiently the novelty because the current politics of the sciences aims to mask the differences beneath hierarchies, to impose a vocabulary that hides the practical heterogeneity of the different scientific fields.

Consequently, one speaks of "Darwinian theory," but is it really a theory in the same sense as, for example, "Newtonian theory"? Why have the American creationists tackled Darwinian evolution and not, as was the case during Galileo's period, astronomy? Because they have judged it as weak. Because they recognized that this science did not resemble the image that is proposed for the sciences. Where is the power to judge a priori, to differentiate, in an episode of evolution, the essential from the anecdotal? In short, where is the relationship of forces between the scientist and his object that every theory claims? Haven't the apparently explanatory grand concepts, adaptation, and survival of the fittest revealed themselves to be empty of explanatory power a priori, simply words that comment on a history *after* it has been reconstituted? And hasn't the temptation to define a priori a general truth of evolution (which, for example, the sociobiologists have given in to) been denounced by the specialists of the field itself?

From my perspective, it is this so-called weakness of the science of evolution that gives it its strength and interest, because this science is not actually endowed with an object, that is, with the power to judge a priori. Quite to the contrary, it has discovered the necessity of putting to work a more and more subtle practice of storytelling. Finished are the monotonous and pathetic histories whose moral agrees so well with our natural judgments. No, the mammals did not conquer the dinosaurs because the dinosaurs were too big and too stupid, an evolutionary dead end, or because the mammals, which lead to us, already displayed the superiority that we are honored with. Contemporary Darwinian accounts no longer have the moralizing monotony that destined the best to triumph. They make continually more heterogeneous elements intervene, which never cease complicating and singularizing the plot that is recounted. Living beings are not *objects of Darwinian representation*, judged in the name of categories that separate the essential from "noise." Each witness, each group of living beings, is now envisaged as having to recount a singular and local history. Here scientists are not judges but investigators; the fictions they propose have the style of detective stories and involve ever more unexpected plots. Darwinian narrators work together, but in the manner of authors whose plots are mutually inciting, learning from each other the possibility of making ever more disparate causes intervene, that is, of giving the status of "cause" to what the moral perspective had defined as "noise." Read the hypotheses that Lynn Margulis dares to propose concerning the origin of sexual difference!

Do the Darwinian plots have a sex? Obviously, and this is the example I gave at the beginning of my presentation, such a plot may be revealed as biased, which is the risk of all fiction. What is important for me is that the risk here can be recognized and is not obscured in the name of the conditions of scientificity of the field. Darwinian evolution does not have an object, but it does not mime science. It has the singular traits that I have attributed to a science: not objectivity guaranteed a priori but the dynamic in which the fictions that we become capable of risk proposing new bonds of interest and meaning between things and humans.

The Darwinian style is an example, but it is not a model. It is important in that it shows that the style of science that was invented, notably in physics, and in other fields where certain phenomena have let themselves be judged and have conferred upon a fiction the power to claim to represent their truth, is only a style, and not the truth of science. Another style, another type of practice is invented when it is a question of exploring the histories in which living beings, continents, soils, and the atmosphere entangle their becomings. What other styles, what other types of risk, interest, and passion will they have to invent in those fields where one currently proceeds *in the name science*? Making this question the *rational* question par excellence is the purpose of putting the right to theory into crisis.

How do we invent the possibilities of an inventive dynamic when it is a question of humans whose singularity lies in the fact that they themselves represent themselves? Neither the judge nor the Darwinian investigator has to confront this annoying relationship of "rivalry" between representations: neither the objects of theory nor the Darwinian histories question the difference of status between the one who invents the questions and produces the testimonies and the one whose testimony is invoked by its representative. How do we invent a positive sense for the fact that humans testify for themselves and are, in principle, capable of being interested in what one says about them, that is, of inventing themselves in the dynamic of the testimonies in which they are at stake?

Let's not be mistaken, I am still talking about the sciences, not posing the ethical problem of the relations between human beings. The sciences present a political problem, but cannot in any way be confused with politics, because it is always a matter of fictions and propositions invented *about* and not *with*. When it is a question of sciences that deal with humans, questions with ethical resonances are always and at the same time technical questions: How can we not mutilate what we are dealing with, unilaterally silence it? How can we constitute the human singularity not as an obstacle but as a challenge for a new style of inventive and corrosive dynamic? Obviously, no one is forced to do science, to accept the distinction that it implies between those men and women who invent questions and search for testimonies, and that about which they work. But, since there are sciences, I think it is worth fighting in order that this distinction be a risk and an invention and not a relationship of forces affirmed in principle.

Will we see one day what is for the moment only an individual state of mind, a painful doubt—am I imposing irrelevant questions, profiting from the position that my culture has given me, from my knowledge and status? Am I judging and silencing without knowing it? Am I becoming no longer a subject of anxiety or of denunciation, but of collective work? For example, will we see strange novels proliferate whose authors would not first of all address themselves to a public but to other author-researchers, novels that would not pride themselves in the solitary genius of creation but whose stakes would be to "bring into history" the question of the ambiguous relation between the passions of the author and those of his subject? I will be told that we would need a novelist's passion that scientists are far from possessing. Where would one find these talented novelists who would, moreover, agree to attempt to make history together? But the passion of doing physics is also singular, and that of the Darwinian narrators is no less so. There is no science without fiction and there is no fiction without passion. Science is not defined by a particular passion that one would call scientific. Each science is born from the fact that the passions that created its field have found the affective, intellectual, and social means of making history together.

Perhaps the perspective that I have opened up is itself a utopia, but I think that it is a good utopia, a utopia capable of complicating the modes of functioning that have succeeded in imposing themselves as normal and inevitable, a utopia capable of weakening what today constitutes one of the sources of power of the established sciences, the fact of convincing the men and women who denounce this power that their position is one of critics, and not of the bearers of new controversies and fictions, a utopia that opposes to the sex of power not another power but the thousand and one sexes of fiction.

Bibliography

Bhaskar, Roy. *The Possibility of Naturalism: A Philosophical Critique of the Contemporary Human Sciences.* Brighton, Sussex: Harvester Press, 1979.

Callon, Michel, ed. *La science et ses réseaux.* Paris: La Découverte, 1989.

Chertok, Léon, and Isabelle Stengers. *A Critique of Psychoanalytic Reason: Hypnosis as a Scientific Problem from Lavoisier to Lacan.* Trans. Martha N. Evans. Stanford, Calif.: Stanford University Press, 1992.

Latour, Bruno. *The Pasteurization of France.* Trans. Alan Sheridan and John Law. Cambridge: Harvard University Press, 1988.

———. *Science in Action: How to Follow Scientists and Engineers through Society.* Milton Keynes: Open University Press, 1987.

Latour, Bruno, and Steve Woolgar. *Laboratory Life: The Construction of Scientific Facts.* Princeton, N.J.: Princeton University Press, 1986.

Lévy-Leblond, Jean-Marc. *L'Esprit de sel.* Paris: Fayard, 1981.

Margulis, Lynn, and Dorion Sagan. *Microcosmos: Four Billion Years of Evolution from Our Microbial Ancestors.* London: Allen and Unwin, 1987.

———. *Origins of Sex: Three Billion Years of Genetic Recombination.* New Haven: Yale University Press, 1986.

Schlanger, Judith. *Penser la bouche pleine.* Paris: Fayard, 1983.

Stengers, Isabelle, ed. *D'une science à l'autre. Les concepts nomades.* Paris: Éditions du Seuil, 1987.

Stengers, Isabelle, and Judith Schlanger. *Les concepts scientifiques. Invention et Pouvoir.* Paris: La Découverte, 1988.

N I N E

Who Is the Author?

Prologue

DURING THE third day of the *Discourse concerning Two New Sciences*, Galileo, in the guise of Salviati, gives a definition of uniformly accelerated movement that he claims agrees with the essence of the naturally accelerated movement of falling bodies. "I say that a movement is equally or uniformly accelerated when, starting from rest, it receives in equal times, equal moments of velocity." For us, this definition is the first statement that falls within the province of modern physics, the first that has stood the test of time in tracing a thread of continuity for more than three and a half centuries. How are the interlocutors that Galileo has created in Salviati, Sagredo, and Simplicio going to react to what for us is an event?

It is Sagredo who speaks: "Although, rationally speaking, I've got nothing against this definition or any other, regardless of the author, since they're all arbitrary, I can nevertheless doubt, and this is said without offense intended, that such a definition, elaborated and accepted in the abstract, fits and is appropriate to the type of accelerated movement that naturally falling bodies obey."

Sagredo is the man of good sense, the one with whom Galileo's readers will identify—a strategy of formidable efficacy, moreover, because when Sagredo, forgetting his presumed impartiality, unites with Salviati, in the *Dialogue on the Two Chief World Systems*, to heap insults on the unfortunate Simplicio and with him all of Galileo's adversaries that Simplicio loosely represents, it is the readers who are, at the same time as him, led to commit a truly moral assassination. The new type of truth invented by Galileo, which has since then been called scientific truth, presents itself openly in the *Dialogue* as a truth of combat, verified by its capacity to silence or ridicule those that contest it.

But in the *Discourse* the tone has changed. Galileo has been condemned. He is an old man who knows that death is near. He writes secretly for readers he will never know. He writes for the future. Simplicio and Sagredo have become simple pawns, asking questions and putting forward objections that Galileo can answer, this being precisely what he considers as the force and novelty of his truth. And Sagredo's objection announces the principal trial that the Galilean truth will have to overcome: to conquer the skepticism with which good sense receives any definition, this one or any other, regardless of the author. The quality of the author, his authority or the rationality of his arguments are no guarantee. Any definition is arbitrary. Any definition, we will say, is a *fiction*, tied to an author.

Sagredo represents common sense. But his objection expresses the fact that common sense is not a transhistorical given, because, during Galileo's period, Sagredo's argument was also the argument of Power, in this case of the Roman church. Monsignor Oreggi, the personal theologian to Pope Urban VIII, has left us an account of the discussion that the pope, then Cardinal Maffeo Barberini, had with Galileo after the first condemnation in 1616.

He asked him if it was beyond the power and wisdom of God to arrange and move the orbits and stars differently, and do this, however, in such a way that all the phenomena that are manifested in the heavens, that everything which is taught concerning the movement of the stars, their order, their situation, their distances, their disposition, could nevertheless be saved. If you wish to declare that God would not be able to do it, you must show, added the holy prelate, that all this could not, without implying contradiction, be obtained by any other system than the one you have conceived; God indeed can do anything that does not involve contradiction.[1]

The great savant, concluded Monsignor Oreggi, remained silent.

That Urban VIII, finding his argument in Simplicio's mouth at the end of the *Dialogue*, considered that Galileo intended thereby to ridicule him, since everything Simplicio says is by definition ridiculous, belongs to the legendary history of Galileo's condemnation, which I will not dwell on. The argument, on the other hand, interests me because it shatters the scenario elaborated by Galileo himself, and which is too often taken up by those who seek to characterize the singularity of what are called the modern sciences. Galileo's adversaries were not simply the outdated heirs of Aristotle, which would have the effect of bracketing the Middle Ages. The truth that Galileo announces needs to do more than simply impose itself on another truth that it would refute. It must first and foremost impose itself against the idea that all general knowledge is essentially a fiction, and that it is not up to the power of human reason to discover the reason for things, whether this relates to the order of Aristotelian causalities or to mathematics.

Barberini, the future Urban VIII, evokes the all-powerful nature of God: "God indeed can do anything that does not involve contradiction." In so doing, he takes up the celebrated argument of Étienne Tempier, bishop of Paris, who in 1277 condemned all the cosmological theses that had resulted from the Aristotelian doctrine. Thus was condemned the proposition according to which "God would not be able to impose a translatory movement upon the heavens, because this movement would produce a vacuum whose existence cannot be admitted without absurdity." An absurdity is not a contradiction. Absurdity relates to the idea of a rationality that would establish, in one way or another, a common meeting ground for human reason and the reasons nature obeys, in such a way that rational argumentation is able to claim the power of distinguishing between the possible and the impossible, the acceptable and the unacceptable, the thinkable and the unthinkable. It is this common ground that is refuted by the reference to the all-powerful nature of the divine author of creation. If God had wanted, what seems normal to us would not be so, what seems inconceivable or miraculous would be the norm. The all-powerful nature of God requires that we think against a background of risk, that we dare, for example, as Samuel Butler did in *Erewhon*, to think that a society could have existed where sickness and misfortune would be severely punished, whereas crimes and misdemeanors would attract pity and the most attentive medical care.

If the will of God is the only difference that can be legitimately invoked between imaginable fictional worlds and our world, any mode of knowledge that cannot be reduced to either logic or pure report is of the order of *fiction*, more or less well constructed. Each fiction can have its advantages, but all of them are tied to an author who is essentially defined by his incapacity to get back to the reasons for the divine decision to create the world like this rather than like that.

If the Greeks had been confronted with the postulation of this all-powerful nature, defined by the absence of constraints, they would have certainly denounced the ugliness of the hubris, of the pride that exceeds all limits, of the despotic decision that draws its glory from its arbitrariness. I will neither discuss here the diverse ways in which philosophers have attempted to return the virtues of wisdom to a despotic God, nor the thorny question of knowing how to relate the history that produces this figure of power in relation to which human reason has to situate itself. For the physicist-philosopher Pierre Duhem, it is the distinctive glory of Christianity to have created, against the certitudes of tradition, a dramatic distance between necessary truths and truths of fact, which it is possible to deny without contradiction. For others,[2] this history is above all that of the commercial towns where, from the end of the Middle Ages, the difference between the possible and the impossible is a matter of will and speculation, where the spirit of enterprise rebels against anything that makes what is and what must be coincide as a matter of principle. Be that as it may, it seems difficult to overestimate the importance of this fact: the Middle Ages created a new figure of skepticism. It is no longer a question of a minority thought, accepting the risk of exclusion or marginality, but of the result of a constraint imposed by power itself, by the church condemning as erroneous from the point of view of faith any use of reason that would limit the absolute liberty of God.

It is on the basis of the connection posed a priori between reason and the production of fiction that I am proposing to tackle the practical singular invention of a new use of reason, a use centered on the question "Who is the author?"

Conquering Skepticism

I am going to propose here a general "definition" of the singular enterprise that is called "the modern sciences." This is not a neutral definition that would give itself the challenge of being equally adequate to anything that now bears the title of science. It involves a singularizing definition, in the sense that its significance depends on the problems that it enables to be asked, on the distinctions that it can give rise to. I will say that, by definition, what are called the modern sciences claim as their starting point that Étienne Tempier and Urban VIII were right. *Normally*, any phenomenon that we observe can "be saved" in multiple ways, each way referring to a human author, his projects, his convictions, and his whims.

From this definition there follows an initial type of literary genre, that of texts that, in advancing an innovative proposition, assert that reason is on their side, that the proposed interpretation must be recognized as rational because it responds to the criteria of a healthy scientific method. These are the texts of "mad scientists." The "mad scientist" advances alone, in most cases armed with facts that should logically deserve general approval. He demands that one take the facts seriously, as epistemological treatises suggest. He becomes indignant, in the name of scientific values, that his proposition is not recognized as scientific. In doing this, he is unaware that if there is a value distinguishing scientific authors from all others, it is that a scientific author cannot claim any value that would require, *as his right,* the interest of his colleagues. *He must invent the means of conquering the skepticism that is always a priori the "normal response"; he must invent the means of having his fiction recognized as not just one fiction among others.*

"Who is the author" of the fiction concerning the movement of bodies that Galileo opposes to Aristotelian science? For it is clearly a question of a fiction, of a world where perfectly round balls take thousands of years to descend a minute difference in level along quasi-horizontally inclined planes, of a world where cannonballs shot from vertically standing cannons appear to follow an oblique course, even within the very barrel of the cannon, of a world where a body goes right through the earth along a tunnel dug expressly for that purpose, a world that, two centuries later, a train launched at 200,000 km/sec will travel over, allowing its passengers to exchange light signals with observers positioned on the embankments.

One can obviously say that this is an abstract, idealized, geometrized, world. The "author" would then be abstraction. But this says nothing new, because one would have simply repeated Sagredo's skeptical objection: it is simply a world answering to a definition elaborated in the abstract. The question is rather to know what has been abstracted, what singularizes this fiction. The fictional world proposed by Galileo is not just the world that Galileo knows how to interrogate, it is a world that no one other than him knows how to interrogate. It is a world whose categories are those of the experimental apparatus that he invented. It is a concrete world in the sense that this world can entertain the multitude of rival fictions concerning the movements of which it is composed and allows for these fictions to be differentiated, to designate which of the rival fictions represents it in a legitimate way. Galileo is not only a writer, although it is highly significant that his activity turned him into a prolific writer. His activity as a writer who distributes roles, arguments, and references is the outcome of the creation of a world capable of witnessing in favor of the fiction that he proposes, capable of guaranteeing that what Galileo says is not only a fiction.

The velocity of Galilean falling bodies is not an abstract notion, the product of an "abstract way of seeing things." Abstract, separable from the moving bodies that it qualifies, was rather the medieval notion of velocity: give me a way of measuring space and time, and you will be able to ignore the difference between the stone that falls, the bird that flies, or the horse who, exhausted and breathless, will soon collapse, and I will tell you their velocity. The velocity of Galilean bodies — the velocity that we would now say defines classical dynamics — is inseparable from the moving bodies that it defines; it *belongs only to Galilean bodies*, to those bodies defined by the existence of an experimental apparatus that, faced with the concrete multitude of rival propositions, allows it to be affirmed that this velocity is not just one way among others of defining the behavior of this body.

Galileo's world seems "abstract" because many things whose categories cannot be defined by the experimental apparatus have been eliminated from it. Scientific "abstraction" expresses first and foremost the fact that victories over skepticism are always local and selective—the stone, but not the bird; laminar flows, but not turbulent flows; definite chemical propositions, but not the catalysis; the germ, but not the "sickness"; and so on—always relative to the invention of a way of discriminating between fictions. This is why the difference between what can be the object of scientific representation and what is supposed to "escape" representation is not of the order of what a theory, philosophical or otherwise, could ground. Grounding always means referring to a criterion that claims to escape from history in order to constitute its norm. Prior to Galileo, who would have held Galilean velocity to be "representable," the instantaneous velocity with which a body travels in no space and no time? Who is it that thinks they can "represent" light, which is neither a wave nor a particle, but that can, according to the circumstances, answer to the representation either of a wave or of a particle? The sciences do not depend on the possibility of representing; they invent possibilities of representing, of constituting a fiction as a legitimate representation of a phenomenon. Scientific "representation" has here a meaning nearer to that which representation has in politics than in a theory of knowledge.

"Abstraction" is the creation of a concrete being, an intertwining of references, capable of silencing the rivals of the person who conceives it. Sagredo is not silenced because he was impressed by Salviati's subjective authority, nor because he was led by some intersubjective practice of rational discussion to recognize the validity of the proposed definition. The experimental apparatus silenced Sagredo, forbade him to oppose another fiction to the one proposed by Salviati, because that was precisely its function: to silence all other fictions. And if, after three and a half centuries, we still teach the laws of Galilean motion and the apparatuses that allow it to be presented—inclined planes and pendulums—it is because up until now no other interpretation has succeeded in undoing the association invented by Galileo between the inclined plane and the behavior of falling bodies.

The scientist who creates an "abstract representation" of a phenomenon, if he falls within *my* definition of science (see who that eliminates), is never alone in his laboratory, like an isolated subject who "would represent" to himself a phenomenon. His laboratory, like his texts and representations, are peopled with references to all those who can put them into question. How does Pasteur represent to himself a microbe? As Bruno Latour writes, "this new microscopic being is both anti-Liebig (ferments are living systems) and anti-Pouchet (they do not breed spontaneously)."[3] Today we have multiplied the modes of intervention of microbes in our knowledges and practices, but the identity of these microbes is still the sum of what the authors have succeeded in making them validate as against other authors. Scientific representation is only abstract in appearance; it is as concrete as the world of discourses and practices within which it intervenes.

To silence skepticism not by invoking some methodological or metaphysical instance that would enable a scientific proposition to be "grounded," but by foreseeing the rival fictions and inventing the apparatus, or the use of the apparatus, that will be able to refute them, such is the invention that both extends to us—we who have more or less forgotten the despotic God of Étienne Tempier—the skepticism that reduces reason to fiction and never ceases to conquer it, by the production, always local, selective, and limited, of ways of discriminating between fictions.

Of course, Galileo claims (Platonic discourse) that the experimental apparatus is simply there to illustrate the truth of the facts, of which he is only the good midwife, which he will lead Sagredo and Simplicio to themselves recognize from the moment they are freed from the illusions of sense or authority induced by tradition. Of course, Lavoisier affirms in the *Méthode de Nomenclature chimique* (1787) that the chemist must rid himself of the imagination that carries him beyond truth, toward fiction, and of all the qualities that would make him an "author," in order that nature can "dictate" the adequate description. Here it is a matter of "methodological" discourses, discourses that present the principles in the name of which the asymmetry between the fiction proposed and the rival fictions should be established. Methodological discourse identifies in what way the rival fictions are recognizable by their errors, prejudices, and naïveté, in short, why they are only fictions. But, as long as the controversy lasts, this methodological discourse has no more authority than the proposition that it afterward appears to ground. After the controversy, if this proposition wins the day, it will make history, methodology will tell why it was normal and rational that the conqueror be victorious, it will have this victory recognized as the straightforward product of scientific rationality since all the rival interpretations will have been recognized as simple fictions.

Methodological discourse is the account of a type of victory that has the singularity of wishing to silence the event that constitutes the victory, the fact that a fiction has found the means of making the difference against rival fictions. Its plausibility stems from the fact that, from the moment an individual scientist knows that he is committing himself to the collective concrete practice we call science, he must formulate his proposition in such a way that it can resist controversy. However, it is not a matter of "ridding oneself of," or of "purifying oneself from," but of integrating within one's own approach the perspective of the polemic that will decide the "scientific" character of the proposition. In other words, a scientist is much more of a strategist than an ascetic. And scientific texts have to be read as strategic works, addressed to potential or already identified adversaries, foreseeing and diverting their attacks, and not as the transparent description of an "objective" approach.

Who is the author? In one form or another, this is the question posed by all scientific controversy. In most cases, it presents itself in the form of a dichotomy: either the scientist can convince his reader-colleagues that he is not the "author" of what he is proposing, that he has restricted himself to taking into account what has been imposed on him *as it would be imposed on anyone in the same conditions of experimentation, observation, and formalization,* or he fails, and finds himself designated as the deliberate (i.e., bad faith) or unintentional (i.e., unlucky) author of his proposition. In the first case, his proposition will legitimately claim "authority," that is, be recognized as "true" in the scientific sense of the term. In the second case, it will be rejected as "unscientific," that is, as a fiction, situated outside of the trials that distribute truth and falsity in the sciences.

Thus one can see in the modern sciences the invention of an original practice of attributing the title of author, playing on two meanings that it opposes: the author, as an individual animated by intentions, projects, and ambitions, and the author acting as authority. This does not involve a matter of naïveté that could, for example, be criticized by contemporary literary theorists, but a rule of the game and an imperative of invention. Every scientist knows that both he and his colleagues are "authors" in the first sense of the term and that this does not matter. What does matter is that his colleagues be constrained to recognize that they cannot turn this title of author into an argument against him, that they cannot localize the flaw that would allow them to affirm that the one who claims "to have made nature speak" has in fact spoken in its place.

Thus the practice of the modern sciences does not presuppose that the scientist can purify himself of what makes him an author. The question is to know if this title of author can be "forgotten," if the statement can be detached from the one who held it and be taken up by others from the moment that they welcome into their laboratory the experimental apparatus whose meaning is given by this detached statement. The question is to know if those who accept this statement can, subsequently, find themselves in the position of independent rivals, or whether the fact that they accept it makes them disciples subjected to the unanimity of an idea.

A scientific statement, if it is finally accepted, is taken to be "objective," that is, no longer speaking of the one who proposed it but of the phenomenon defined as remaining open to other inquiry. Thus it is supposed to refer not to the author but to "nature" as the *authority*, as if it was clearly nature that, following Lavoisier's expression, had dictated its "truth." There is little point in asking if this "truth" should be thought of as a "reflection," a "representation," or an "expression" of reality. The "truth" in the scientific sense does not answer to a positive criterion whereby one could recognize it. It indicates a historical landmark, the recognition that an author's fiction has succeeded, against a background of a priori skepticism, in being taken as authoritative.

History

Authority and author have the same root and medieval scholastic practices gave them an interdependent sense. Authors, in the medieval sense, are those whose texts act as the authority; they can be commented on but not contradicted. However, this in no way implies a submissive reading—rather the opposite. Thus, in Saint Thomas Aquinas's *Summa theologica*, the authors are called to witness on a specific question in the form of citations abstracted from their context. The game and the stakes are to make them come to agreement by taking account primarily of the letter of the citation, not by discussing the meaning given to it by the author. In other words, the author acts as the "authority," but Saint Thomas acts as the judge and treats the author-authority as a witness called to appear: he must assume that the witness has spoken the truth, and his judgment will have to take into account this testimony, but it is Thomas who will actively decide in what way this testimony will be taken into account.

In the same way, every scientist knows that each phenomenon studied can generally be interpreted in many ways. This also is of little consequence. What does matter, and what is at stake in the difference between a fiction with a scientific vocation and a proposition recognized as scientific, is the active invention of ways of constituting the world that is under interrogation, as a reliable witness, as a guarantor for the one who speaks in its name.

The difference between scholastic and scientific practice is thus not as radical as one might have thought. Saint Thomas recognized that the "authors" acted as the authority, but he behaves as if he knows that he is free to determine the manner in which they must be taken into account. Scientists recognize "nature" (the phenomena that they are dealing with) as the only "authority," but they know that the possibility for this "authority" to act as the authority is not given. They have to constitute nature as the authority.

The principal difference concerns the relation between authority and history. The scholastics attempt to put the authors—pagan philosophers, Christian doctors, and the divine Author of creation—in agreement. Their goal is to stabilize and harmonize history. As far as science is concerned, succeeding in constituting nature as the authority and making history are synonymous. The scientist, as an author, is not addressing himself to readers, but to other authors; he does not seek to create a final truth, but to create a difference in the work of his "author-readers." And it is in terms of this difference, in terms of the risks its acceptance entails and the possibilities for new problems it opens, that a statement is evaluated and put to the test. The scientist does not seek a stable accord; he endeavors to constrain history by what he does; he endeavors to ensure that history is forced to go by way of the testimony of the things that he succeeded in giving rise to, by way of the difference between fictions and truth that he succeeded in creating.

The sciences often give the impression of an "ahistorical" enterprise. If Beethoven had died in the cradle, his symphonies would never have seen the light of day; if Newton had died at the age of fifteen, someone else would have taken his place. This difference relates in part to the stability of certain problems, for example, to the empiricomathematical ellipses of Kepler, which were obviously capable of persisting as an issue. But this difference is not as general as one might think. Thus, I think that I could claim that if Carnot had died in the cradle, thermodynamics would not be what it is today. In any case, the more or less stable persistence of problems is not enough to explain the ahistorical appearance of the sciences. It must be added that this appearance expresses the intensely historical character of this enterprise, the relation between truth and history. Innovative scientists are not subjected to a history that would determine their degrees of freedom; on the contrary, they take the risk of inscribing themselves in a history and of attempting to transform it, but of transforming it in such a way that their colleagues and those who, after them, recount the history will be constrained to speak of their invention as a "discovery" that others could have made. The history of the sciences does not have as actors, humans "at the service of truth," if this truth is to be defined by criteria that avoid history, but rather by humans "at the service of history," whose problem is history, and the truth, here, is that which succeeds in making history.

A strange history, then, that associates human arguments and the testimonies of things in continually renewed ways. "The order of things," one might say, with the difference that it is not a matter here of "representations" in the sense that this history would be a purely human one concerning a "reality" defined as silent, but of the attempt—involving mainly costly methods, mobilizing laboratories, particle accelerators, observatories, studies, collections, expeditions—to invent ways according to which this "reality" will be recognized as having the power to intervene among human authors, to "make a difference" between their arguments.

The history of the sciences is thus not only the history of the controversies that decide the undecidable question "Who is the author?"—the world or the human that interrogates it. It is also a history in which are invented the responses to another sense of the question "Who is the author?" Because "Who is the author?" can also mean "What passions and risks would define the author?" or again, "What genre would the fictions capable of making history come under?"

I propose to consider that the creation of the Galilean object corresponded to the invention of the author Galileo, that the creation of the experimental sciences corresponds to a type of author, both poet and judge, who did not exist prior to this creation.

Poets and Judges

What is a poet? Etymologically, a fabricator. The history of language has changed the meaning of the term, but has also dramatized one of the implications of the art of fabrication: the dimension of creation that does not refer to anything but itself and that is not accountable to anything but itself. Poets, then, are those who gives themselves the freedom, and take the risk, to invent and bring into existence that which they speak of.

What is a judge? In the most general sense, it is someone who acts "in the name of"—in the name of the law, of course, when it is a question of the legal code, but it is not just a question of the code. The judge exists from the moment when that in the name of which he speaks and acts authorizes him to determine what, in a concrete situation, is significant and has to be taken into account, and what is secondary, a simple parasitic noise that can be "abstracted," in actuality or intellectually eliminated. The judge is the one who knows, a priori, according to what categories it is appropriate to interrogate and understand that with which he is dealing.

The experimental approach, as it was invented in the seventeenth century, thus turns the scientific author into a singular hybrid between judge and poet. But the term "approach" is a dangerous one. It implies that experimentation is a general principle that can be pertinently applied in all cases. Rather than speaking of approach, one should speak of the *event* that constituted, and that has subsequently continued to constitute, on every occasion that it occurs, the practical discovery of the *possibility* of submitting a phenomenon to experimentation.

Submitting a phenomenon to experimentation is to actively produce it, to "re-create" it, and have it accepted that this re-creation is simply a "purification," restricted to eliminating "parasitic effects" in a manner that makes the phenomenon capable of speaking its truth. The scientist-poet "creates" his object; he fabricates a reality that does not exist as such in the world but is rather on the order of a fiction. The scientist-judge succeeds in having his creation accepted as a discovery, and that the reality he has fabricated testifies that the hypotheses in the name of which it was created are precisely those that render intelligible the "natural" phenomenon studied. The poet's creative power must be differentiated from the law of the strongest (and, since Étienne Tempier, we have accepted that no phenomenon is naturally "stronger" than our capacity to create fictions). Experimental creation has to gain the recognition that it is nothing but a legitimate purification that gives a natural phenomenon the power of testifying in favor of the experimenter. This is why the failure that the experimenter most fears is that of the *artifact*, the "fact of art": he really has "created" an experimental reality, but this reality has been recognized as incapable of testifying in relation to the natural phenomenon; his creation has not been able to win the title of "simple purification." He has created a laboratory fiction.

The literary genre to which experimental literature best corresponds is the epic or initiatory genre: a hero, through the many trials during the course of which he is transformed, succeeds in acquiring a treasure, a secret, an answer. Who is the hero? It is neither the scientist nor the phenomenon; it is the bond that unites them; it is what the interrogated phenomenon allows the scientist to claim in its name. What are the trials? They are the questions, objections, and possibilities for controversy that are more or less explicitly presented within the text and then thwarted. Who is transformed? Neither the scientist nor the "natural" phenomenon for which the quest was undertaken, but the experimental apparatus that becomes capable of giving a sense, from its own point of view, to that which menaced it, and of putting an end to the menace, or better, on occasion, of converting it into an argument in favor of the apparatus's creator. The worst enemy becomes an ally. What is the object of the quest? The power of being able to say "nature has spoken," when confronted by colleagues that one presumes have now been reduced to silence.[4]

The art of the experimenter is in league with power: *the invention of the power to confer on things the power of conferring on the experimenter the power of speaking in their name.* In most cases, the event constituted by the conquest of this power is concealed, in public texts, behind the calm description of a set of experiences that lead naturally to a rational conclusion. But, for the readers to whom it is addressed, this text is far from being "cold"; it is a risky device that brings together, at the same time and indissociably, the "facts" and the readers, and proposes roles for them — pertinent critique, incontestable authority, ally, unsuccessful rival — that they will either have to accept or invent the means of refusing.

In general, when an experimental fact is accepted, and often in the very process of its acceptance, a new question, a new history begins: What does this fact testify to? Who will have to take account of it? Who will be affected by the new constraint that it constitutes? In short, what is its significance, that is, what history is it the bearer of? Sometimes, this question presents itself with a discretion that signals great confidence. Thus, at the end of the paper published in *Nature* where they announce the double-helix structure of what we now call DNA, Watson and Crick remark that "it has not escaped our notice that the specific pairing we have postulated immediately suggests a possible copying mechanism for the genetic material." This short phrase calmly announces a shattering event to all its readers. The reconstitution, from the sum of the available experimental data, of the structure of a macromolecule that is found in every cell was obviously something important, and perhaps even deserved the Nobel prize. But the double helix is not like other molecular structures. It seems "to speak for itself," "to immediately suggest" a solution to the biologist's old problem: how, when two cells divide, can each possess the totality of the genetic heritage of its species? And what is more, in this solution, a molecule capable of producing two copies of itself is of a technical type, as if living beings had, prior to man, discovered the answer to the problem of the transmission of memory in a text that can be copied, letter for letter, without understanding it. Watson and Crick know that the consequences of the "fact" that they report are incalculable; thus, they can content themselves with a "short phrase" that now figures in the rhetorical history of the sciences as the most accomplished example of a litote.

However, in most cases, a "fact" is not in and of itself so talkative. Its significance, as well as its recognition, involve a history that is produced by active strategies. Whom will it "interest"—that is, who will agree to associate his research program with it, to be situated by it, that is, let it "be between" (*inter-esse*) his own questions and those that produced it? This is a crucial question, because what we have to call the creation of a reality depends on it. Indeed, reality is of course not what exists independently of human beings, but that which demonstrates its existence by bringing together a multiplicity of disparate interests and practices. Is the big bang a reality? If one can still be unsure about this question, it is not because it is unobservable, or because it corresponds to an unrepresentable singularity, but because, up to now, this consequence of relativistic cosmology has not made a difference for other sciences, has not succeeded in proposing to other fields the hypothesis of fictions that would make them mutually supportive. Do atoms, or microbes, or genetic material, exist? Yes, affirms a crowd of academic, medical, and industrial laboratories, a crowd of diverse theorists, but also philosophers or specialists in ethical issues; because, if they do not exist, we do not exist either, we who have been produced by a history that only their actual existence can explain. The primary characteristic of that which we say "really exists" is to be the stable reference of many disparate practices. But this stability is itself constructed. If a large number of disparate practices have succeeded in constructing themselves around a common reference, this reference is stable in separate relation to each of them.

Large-Scale Maneuvers

With the double helix, nature spoke of genetic material, surely, but did it say that from now on genetic material constituted the key to living beings? Did it tell embryologists, for example, that they had to reformulate their questions in a way that took genetic material as the starting point, that they had to understand the development of the living by taking bacteria as their model? Remember, bacteria are the reliable witnesses to the key role of genetic material precisely because they do not develop. Yes! claimed those one calls "molecular biologists." No! protested the scandalized embryologists. Well, maybe, hesitated the institutions that finance research, the young researchers, the popularizing journalists, the editors of scientific manuals— all those whose choices, at this level, construct history. The texts are no longer addressed solely to reader-author colleagues. They are now addressed to readers who are asked to take into account the consequences of a history that is not of their own making, of which they are not the authors. Correlatively, the status of the authors also changes. They now speak in the name of their discipline. From now on, it is molecular biology, high-energy physics, or neuronal science that makes promises, claims, and demands.

We are dealing with texts that one might be tempted to read according to the distinction "true science"/"ideology"—an erroneous reading if one thereby understands that ideology constitutes a kind of parasite from which the sciences should or could be purified; a justified reading if those who denounce "ideology" realize that in so doing they are *engaged* in a history that concerns them because it is a history whose function is not to affirm a local and selective triumph over skepticism but a global reorganization of knowledges and human practices. In any case, it is an ambiguous reading, because the denunciation is made "in the name of the right"—the right to separate the "truly scientific" from what is not—while those to whom it is addressed are in the process of fabricating this right.

The history that is constructed here will determine, on a more or less large scale, the modification of the conceptual landscape in the most concrete sense of the term: research programs, funding, prestige, the recruitment of young people attracted to the leading disciplines. It undertakes to reorganize the questions, and those who ask them, according to a hierarchy in which it obviously constitutes the summit: essential questions, corresponding to fundamental research, or to "applied," hierarchically secondary questions, parasitical questions, obscurantist, false, questions to be abandoned, and so on.

The authors are still judges here, but they no longer judge things without at the same time judging knowledges and those who put them into action. They are still poets, in the sense of fabricating, but they do not simply fabricate links between words and things, reliable witnesses capable of intervening in discussions between colleagues; they fabricate *concepts* that, in the name of the created link, have to produce a transformation of the conceptual landscape, and above all of the qualification of the authors who work in it.

Indeed, a conceptual landscape, at any given period, qualifies these authors. Moreover, those qualifications involve social definitions. Thus, since the beginning of the nineteenth century, it has been accepted that engineering science is defined as "only" an application of physical theories, and therefore cannot question them. Engineers have known for a long time of the phenomena of nonlinearity that now enthral mathematicians, physicists, and chemists. But they did not form part of those who "make nature speak." Other "qualifications" may be related to quite precise disciplinary claims. Thus, with *Neuronal Man*, Jean-Pierre Changeux, in the name of neuronal science, proposed that all those who work in the domain of the sciences called "human" be qualified as "lacking," that is to say, "waiting for," this "modern biology of the mind" that neuronal science will one day produce.[5] Yet others imply what is called a "conception of the world," that is, a reference to the "fundamental truth" of nature as a principle for the hierarchization of the sciences. Thus, since the beginning of the century, the "probabilistic" interpretation of physicochemical irreversibility qualifies both physicochemical phenomena and those who study them: what really accounts for the difference between these phenomena and the reality described by the fundamental laws of physics is not nature but the incapacity of humans to observe and describe these phenomena according to those laws. As a result, the physicochemist is judged by his divergence from the ideal that would be constituted by "Maxwell's demon," which, for its part, is capable of describing not average populations of particles, but each particle individually.

The texts that involve "large-scale maneuvers" where the landscape of the sciences is modeled do not speak about what is, but about what will be, or about what should be. They do not announce the event that is constituted by a local and limited victory in relation to the undecidability of fictions. They announce a fiction that would allow the judgment in principle of what they do not have the means to attain in fact. They announce that, depending on the case, these means will one day exist, or that they are beyond human capacity but nevertheless allow these capacities to be qualified. At any rate, these texts belong to a *prophetic genre*, pronouncing in relation to what is "in the name" of instances that, in one way or another, transcend what is. And here, the controversies cannot be determined in laboratories, because at stake is the significance of what goes on in these laboratories, the kind of judgment that the laboratory authorizes to be taken outside of the laboratory. It is then that arguments appear, bearing more or less explicitly either on scientific rationality or on what the history of the sciences has taught us, or on what we can and must wish for man, in terms of economic, industrial, medical, and historical development, or on the "true reality" beyond deceptive appearances.

It still needs to be emphasized that the "prophetic genre" is not a parasite on the history of the sciences. The technicoscientific "reality" that is the given we have inherited was produced by judges-poets-prophets. Ideology, as a category designating the past, designates prophets who did not succeed in making history, in making us their heirs. As a category active in the present, it designates a position inside a controversy that excludes outside, "neutral" characterization.

Other Styles

Judges, poets, and prophets are terms that designate styles as much as talents and passions. Of course, these styles, talents, and passions preexisted the invention of the sciences, but the event constituted by the discovery of the possibility of theoreticoexperimental sciences constituted them as scientific — as scientific, but not for all that, as defining the sciences.

The definition of the sciences that is my starting point is, as I have said, singularizing. It involves a certain manner of speaking about the sciences, and, for example, of posing the problem of those texts that can be called scientific although they only mimick the juridical style from the experimental sciences. Here, the authors become judges, without, however, being poets — in the name of the method, in the name of objectivity, in the name of the right that every science would have to "define its object." My definition requires me to emphasize that we are here dealing with boring texts, in which no risks are taken, but in which the statistics are generally impeccable because it is the method that is target of the trials.

My definition also engages my interest in the opening up of new types of history, in the proliferation of new links between science and fiction that complicate the question "Who is the author?" I am thinking in particular of computer simulations that are transforming the status of what one calls a model. For a long time the "model" has signified, in scientific practices, a "poetic" activity that is incapable of making its author a judge. The model has an author who knows he cannot claim to be forgotten. But the model simulated on a computer introduces such a distance between the author's hypotheses and the engendered behavior that the author speaks of the computer in the same way as the experimenter speaks of the phenomenon: as if, adequately interrogated, it can act as the authority. "The simulation shows that..." is a statement that, henceforth, sometimes takes the role of "the experiment shows that...." Here there are new histories beginning, with new types of authors, new stakes, and new controversies.

But here I would like to talk about a question in relation to which my definition of the sciences engages me much more actively and which is without doubt actually at the origin of this definition. The question concerns the limits of the experimental model created by the poets-judges-prophets. If it is the case that every science (save mathematics and logic, to which Étienne Tempier's God remained subjected) attempts to make "nature" intervene in man's arguments, nothing says that it must always confer the power of a judge on those who actively seek to represent it. All experimentation depends on the invention of a relationship of forces that enables the creation to call itself a simple purification and that enables the poet to call himself a judge. But it is not within the power of human beings to decide to make this relationship of forces exist. The question that interests me is that of the scientific styles that can be invented, not, as is the case with simulation, through the existence of a new technical apparatus, but through the positive, practical invention of scientific authors who address themselves to nature without waiting for it to confer on them the power of judging.

For some years, we have been assisting not in the birth but in the positive invention of what I will call "Darwinian authors." The distinction between birth and invention indicates that Darwin is now recognized not only as the founder of the science of the evolution of the species, but as the "first Darwinian author," the creator of a style much more than an explanatory hypothesis.

The history of this retroactive invention involves quite different episodes. We'll look at three of them.

The first is the resurgence of the creationist controversy in the United States, a process in which the Darwinians have had to affirm their status as scientific authors when confronted with adversaries who emphasized the difference between their practice and that of authors who act like judges and poets: "Darwinian theory" does not offer the means of judging a priori what, in a given situation, is important or simply anecdotal; the "concepts" of adaptation or of survival of the fittest do not have the power to enable the scientist to anticipate, in such a situation, the manner in which they will apply. The second is the history that issues from Alvarez's hypothesis according to which the disappearance of the dinosaurs was linked to the impact of a giant meteor. It brought up the scenario of a "nuclear winter," the effects, similar to those of a meteor impact, that would be produced by a nuclear war, and it continues today with the problem of the "greenhouse effect," which addresses a dynamic of effects centered around the light and calorific exchanges that singularize the earth. The third episode relates, strictly speaking, to evolutionary biology, and to the transformation of its connections with the paleontological data. Stephen J. Gould and Niles Eldredge's theory of "punctuated equilibrium" has brought into question the power of the idea of continuous, gradual selection, according to which the paleontological data, which are marked by discontinuities, should be judged incomplete and misleading. Correlatively, the "adaptionist" narrative has lost the status of being a unique narrative. The narrative mode becomes centered around a problem, specifically that of the question of the stability or instability of specific identity.

These three episodes have given rise to authors whose common characteristic is that they have abandoned the risks of being judge for that of narrator (the models of nuclear winter and the "greenhouse effect" are narrative models). One could go so far as to say that Stephen J. Gould's *Wonderful Life*[6] constitutes a manifesto somewhat analogous to the *Dialogue on the Two Chief World Systems*, not, of course, because Gould would be a "new Galileo," but because it heralds the invention of a new style of author characterized by a new type of "ethos." Gould is not a prophet in the sense that I have defined; he does not speak in the name of a power to be recognized. He communicates the new type of passion, risk, and interest that singularizes Darwinian authors: the invention of earth and the living beings that inhabit it as witnesses to a long and slow history.

The history of the texts stemming from the Darwinian tradition disappoints all those who expect that a science will be able to be recognized by its capacity to silence those who criticize it. Right from the beginning it was put to the test: how will blind selection ever be able to explain the invention of an organ as complex as the eye! And, in fact, "evolutionists" still cannot quite explain how an eye was created, but they have succeeded in "making history" with living beings in a manner that reinvents the way we look at them, which has multiplied the questions. They have interested us in the curious traits that indicate what long duration is capable of rather than on the appropriateness of the eye to vision. Whereas, since Aristotle, we perceived everywhere the logic of relations between ends and means, the "Darwinians" have succeeded in interesting themselves, and in interesting many of us, in the strangeness of the "panda's thumb,"[7] or of turtles crossing the Atlantic in order to reproduce. They have not succeeded in explaining living beings, but in constituting them as witnesses to a history, in understanding them as recounting a history whose interest lies in the fact that one does not know a priori what history it is a question of.

The Darwinian genre has similarities with that of "whodunits": how can one explain *this* type of behavior, *this* anatomical form, *this* mode of reproduction? Each of these explanations is local: none of them confers on the author the power to silence other authors, who investigate other intriguing traits. And yet, the Darwinian authors have managed "to make history together," that is, to make the testimony of one domain intervene in the description of others. Here, what makes history is marked by the *singularization* of histories. Darwinian authors learn vigilance from each other, the necessity of exploring the diversity of causes and the diversity of ways in which the same cause can cause. They learn distrust in relation to any cause that might carry with it the claim to determine how it causes, and this distrust is, correlatively, identifying a more general pitfall: the diverse modes of assimilating history to progress. In *Wonderful Life*, the "role" of Simplicio is held by "our habits of thinking," which always tend to define what happened in terms of what had to happen.

Darwinian authors are thus neither judges, poets, nor prophets, because the history of life as they have learned to read it does not authorize any principle of economy, that is, it does not permit the invention of the relationship of forces that would allow an object to be created and judged, or a hierarchy of questions to be established. But Darwinian authors nevertheless rely on a relationship of forces. Their questions presuppose and imply the stability of the difference between the present to which they address their questions and the past that they attempt to recount. This difference finds an analog in fictional genres: for example, the distinctive characteristic of the classic detective story is that the difference between the police investigator and the suspects is stable. The crime, if it happened, took place *before* the intervention of the investigator. The rule of the genre in Darwinian narratives is of the same type: the traits that interest them are the product of a long history and thus have a stable identity in relation to the type of intervention that enables them to be studied.

The situation of the scientific author is quite different when (for example) the rats, baboons, or humans that he is dealing with *are capable of interesting themselves in the questions that are asked of them*, that is, of interpreting from their own point of view the sense of the apparatus that is interrogating them, of existing in a manner that actively integrates the question. The situation is quite different when the history whereby the one who interrogates seeks to become an author *also makes a history* for that which is interrogated, that is, when the conditions of production of knowledge for the one are equally and inevitably the conditions of *production of existence* for the other. Here, the notion of witness becomes ambiguous: notably with humans, the scientist is dealing with beings who are capable of obeying him, of attempting to satisfy him, of agreeing, in the name of science, to reply to questions that are without interest as if they were relevant, indeed, even allowing themselves to be persuaded that they are interesting, since the scientist "knows best."

Is it necessary, in order to remind ourselves that here science and ethics are indissolubly linked, to recall the experiment in which Stanley Milgram, in the name of psychological science, created the conditions under which normal individuals would become torturers? Is it also necessary, in order to remind ourselves that here the ethical question is always also a technical question, to emphasize that Milgram's experiment did not produce reliable witnesses? It did not confer any authority to a particular statement, but rather reproduced, in an experimental setting, the perplexity that human history constrains us to. Milgram's torturer-subjects knew they were at the service of science, and this knowledge had as a consequence that the experiment, which was supposed to restrict itself to bringing a behavior to light, without doubt contributed, in an uncontrollable way, to producing this behavior. If a living being is capable of learning, which is also to say of defining itself in relation to a situation, the protocol that aims to constitute this living being as a reliable witness in the experimental mode and thereby constrain it to reply in a univocal way to a question decided by the experimenter creates an artifact.

When that which is interrogated cannot be defined as indifferent to the manner in which it is interrogated, when it learns, that is, changes, according to the manner in which one interrogates it, the scientific author himself is exposed to new types of risks that I will call *pathetic.* In fact, to the extent that what he is dealing with can no longer act as the "authority," he can no longer allow it to be forgotten, or forget, that he takes part and bears some responsibility for what he describes. A scientist is now the one who takes the initiative of posing a question, and the "pathetic risk" is the risk of *bearing* the backlashes of the initiative, of facing the question "Who am I for the other?" These backlashes may be strictly physical, as was the case with the unfortunate ethologist who was attacked by a chamois: he thought he was "observing without intervening," but, from the point of view of the chamois, his neutrality signified "domination." Without knowing it, he had overcome a series of visual challenges and was thus the "dominant male," until the day when a rival chamois chose to pass to the next stage of the confrontation. The backlash can be affective: the ethologist Shirley Strum, an exception among ethologists, has recounted her discovery that to understand baboons was also to take the risk of loving them, that is to say, of being transformed by them.[8] The backlash can bring into question the very quality of being a scientific author: how does one study a different human group without putting one's own social and cultural identity at risk? Ethnologists, who speak of the temptation to "go native," have always known this risk of experiencing the emptiness of the project of "reporting" what has been learned to those who are, from then on, no longer "colleagues."

These "occupational hazards" define the singularity of "material" that has not yet really gained recognition for its "literary genre." Today they correspond mainly to distressing feelings, carefully concealed in the name of the method, subjects of anxiety or of denunciation. Yet they designate, in the negative, the style of author that this "material" could create if one day it actually becomes "scientific material" according to the definition of science that I have given. Will we see scientific authors trying to "make a reality speak" that they know is engaged with them in a common becoming? Will we see authors learning from each other the risks of fictions that continually make the sense and implications of their own initiatives more open to question? It will be said that it would require a passion for lucidity that scientists as we know them are far from possessing. But the passion for doing physics is also singular, and that of the Darwinian narrators no less so. There is no science without fiction, and there is no fiction without passion. There is no "scientific passion" but the invention, always singular, of a "scientific becoming" of passions.

III

PART

Science and Society

T　　　　　　　　E　　　　　　　　N

Time and Representation

(with Didier Gille)

ON JULY 27, 1852, the Brussels town council finds itself confronted with a dilemma: either to yield to aesthetic factors or to surrender to technical reason.

For some years the facade of the town hall has been undergoing restoration. On the central tower, a clock dial, installed twenty-three years earlier, gives the official time. Its mechanism is extremely precise, but unfortunately the dial itself badly obscures from view certain architectural features.

Should the clock be moved to another monument? But it seems inconceivable that the town should lose the control of time. The burgomaster opposes the idea vigorously in these terms: "I attach great importance to the regulator of time being at the town hall rather than at a church or at another monument that isn't under the direct control of the local authority."[1]

So, what about somewhere else in the town hall? But in this case the precision of the clock would have to be sacrificed as the dimensions of the current mechanism prevent it from entering the corner turret that is destined to support the clock dial. That year, the situation remains unresolved.

Six months later, a deputy burgomaster suggests a quite new technical possibility: the electrical transmission of impulses between the mechanism and the clock dial.[2]

From now on, there is nothing to prevent the clock dial from being moved while the mechanism remains in the central tower.

What is more, a new possibility immediately opens up: the indefinite proliferation of electric clocks giving the official time all over town, even in private houses.

In December 1856, the decision is taken to install, throughout the whole town, a network of conducting wires to which will be attached one hundred public clock dials. Moreover, the principle of a subscription for individual houses is accepted and will be fully implemented in subsequent years.[3]

Laws and Norms

In the same year, 1856, another question troubles the town council: should the curfew clock [*cloche de retraite*] be taken away?

The last relic from the time when the Joufvrouwencloke, Werckcloke, and Lestecloke marked out and regulated civilian life, this bell rang the closing time for the taverns and theaters. Its removal is envisaged as the end of an anachronism.

However, behind this proclaimed obsolescence, a reading of the council bulletin reveals altogether different stakes. Apparently, the ordinance is quite simple: it stipulates that public establishments will have to close at midnight. But something fundamental has changed: there will no longer be a sound signal to dictate the conduct of those who are out late. It is up to them to know what time it is.

Previously, the sound of the bell told everyone the law at the precise moment when it had to be known. Now, the law is proclaimed once for everyone, and each person must be individually responsible for the conditions of its actualization; remember that this is precisely the moment when it is decided that the town will soon be populated with electric clocks taking their time from the central clock, often designated as the "legal regulator of time."

But that is not enough. This regulator, even when multiplied, is silent. What will have to occur simultaneously is the establishment of an infrastructure and the institution of a habit, a conduct: people will have to ask themselves what time it is. The operation that consists in replacing a regulation that must be continually repeated with a law that is stated once and for all seems to us to characterize the passage from a legal space proper to the ancien régime to normalized modern space. The constraint no longer bears on the physical body but on the body as supporting behaviors. The constraint no longer occurs strictly in the mode of prohibition but is from now on accompanied by a positive incitement: the putting into place of material infrastructures on the basis of which "normal" and "natural" models of behavior will be able to be internalized.

To the omnipresence of the law silently produced by clocks corresponds the normalization of conducts; thus it is not surprising that the use of the *cloche de retraite* should be described as "abnormal" in the discussions of the local council.

This same local council will discover that the normalization of conduct *transforms* the social space in which the law is applied. If the civil law is modeled on the inexorable law of time, it logically follows that any space of negotiation should be absent.

We know that the moment at which the bells rang was, to some extent negotiable. The convention was not totally beyond the control of those who were subjected to it.

On the other hand, when the law proclaims the hour, the working out of tolerances will have to be done in an arbitrary, tacit fashion, as is demonstrated by the following exchange:

VANDERLINDEN:

I would like the law to determine in which cases establishments could stay open after the normal hour.

THE BURGOMASTER:

This is how I will use this arbitrary power: I will never give authorization to establishments that are suspect, where I know that the extended opening hours would only be used for activities that would undermine morals or public order. This is a judgment that cannot be put into rules.[4]

The Autonomization of Social Time

One of the dimensions of the history that we have just described is the autonomization of the law and the consequent transformation of the space of social negotiation. We will find this same process, which is both social and technical, in the problematic of the unification of time during the nineteenth century.

Right from the start, the introduction of mechanical devices for measuring time had posed the problem of the relationship between astronomical time and clock time. The first mechanical clocks, as is also the case with the sundials that have remained from antiquity, are not timekeepers but ways of representing and marking the course of the sun. The base unit was the diurnal period, divided into twelve hours of equal length, which involves a variation in the length of the hour according to the time of year. Thus, the first clocks were not subjected to any constraint pertaining to precision and regularity; on the contrary, the mechanism had to provide for a variability in the speed of the clock hands so that they could be adjusted to the variation in the length of the "temporary hours."

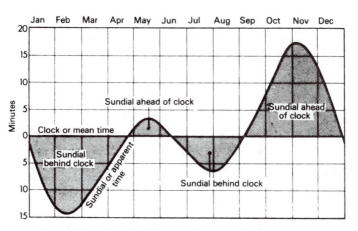

Fig. 1. "Time Equation" showing the interval between solar time
and clock time during the course of the year. *Source:* Derek
Howse, *Greenwich Time* (Oxford: Oxford University Press, 1980).

The adoption of equinoctial hours, being of equal length throughout the year, occurred during the fifteenth century but nevertheless did not bring
about the autonomous working of clocks. It is a fact that the moment when the sun is
at the zenith does not correspond to midday on the clock: the sun is periodically in advance or behind in relation to the clock (see figure 1). This difference was corrected
every day: one "jumped" the clock hands in order to "keep the clock on the sun."[5]

For the first time, in 1780, in Geneva, an average time was
adopted that creates a seasonal difference between the meridian midday and midday
on the clock. For the first time, instead of following solar time by way of daily approximations, social time cuts through it transversally. The now-uniform law becomes autonomous in relation to the natural course of things. Such an event did
not go by without dismal prognostications:

*The sun's midday will no longer fall in the middle of the day, it was said, at twelve noon by the
clock. Tradesmen and day laborers will be confused at work. The morning will sometimes be
longer and sometimes shorter than the afternoon. Bakers, misled by clocks, will no longer
be ready on time, and the population will go without bread.*[6]

The social body itself seems impressed by the boldness of its
rupture with the order of things.

Let's now turn to the beginning of the nineteenth century. In
Belgium, during this period, each town had its own local time, based more or less
on the course of the sun. All that will change with the arrival of the railroad.

The necessity of having the same time for the whole network poses the question of a common system of reference. The famous Belgian astronomer Adolphe Quételet will deal with this problem: he will trace meridians across the whole of Belgium so that at each point the train time—which is, moreover, the time in Brussels—can be deduced from the longitudinal position of the point and its average solar time.

Beginning in 1845, this enormous undertaking will be made redundant by the telegraph, whose lines precisely follow those of the railroad. From now on, the time in Brussels can be instantaneously transmitted to every station.

Progressively, one will see the uncoupling of local time from solar time and its redefinition as a function of railroad time.

Each locality continues to have its individual local time, but from now on the only reason for this time is to help travelers; it helps them to avoid missing the train by being a few minutes ahead of the station time. From then on, whether the towns be situated to the east or to the west of Brussels—that is, whether their solar time is ahead of or behind the time at Brussels—their local time will be systematically ahead of Brussels.

This has a baroque effect since, uncoupled from solar time, local times now proliferate in an arbitrary and delocalized manner.

On May 1, 1882, Belgium adopts a unique legal time, that of the Greenwich meridian, and does this on the advice of its minister of railways.

From now on, the growing density of international circulation imposes a coherence between the different times of the national railways: the system of time zones is established.

The rupture between legal time and solar time is complete. Even Brussels uncouples from the sun.

It is, moreover, within the Brussels town council that the most strident opposition will occur. The burgomaster declares: "If the railway authority demands the adoption of a universal time, or at the least of a conventionally regulated time, the civil authority demands that the adopted time correspond as much as possible to the real subdivision of time, that is, to the course of the sun."[7]

Another adversary of the reform goes even further:

Every day new railways penetrate into the midst of our rural populations, who hold onto their customs and know only of the solar time whereby they regulate their work. To take this away from them and replace it with a foreign country's time would be, in my opinion, an injudicious measure. It would needlessly frustrate the customs of the whole country.[8]

This argument has to be admired. The penetration of the railways is presented as if it has no effect on the habits of the population, whereas the time change would produce destructive effects. The argument plays on the symbolic force of the rupture. If the railroad still corresponds to a process of negotiation with nature and with the social, Greenwich time, from the moment it is proclaimed, is autonomized from nature and the social.

This situation of rupture is presented as a confrontation between scientific rationality and the lived world. Thus the burgomaster of Brussels warns: "By transporting the bases of the measurement of time into the domain of abstraction, one will perhaps have gained control over the procedure used, but its character will be changed. It will no longer be the natural expression of a lived reality, but will become an arithmetic or administrative expedient foreign to the most general public interests."[9]

Metaphors

The scenario of the opposition natural language/scientific language, lived reality/arithmetic expedient is an interesting case of displacement of the stakes.

It is clearly economic, political, and even military forces that imposed a redefinition of time enabling the control of exchange and communication, an instantaneous location of events happening anywhere in the world.[10] Thus, it is not an opposition of the type social life/scientific convention that is at work here, but rather a set of contradictions deployed within the social field.

Obviously, a very real fear is at work among the Brussels bourgeoisie and is manifested in its reservations: by virtue of the circulation of human beings, goods, and information, Belgium begins to escape from its control and now becomes part of the world system.

Be that as it may, and although the argument only serves to obscure the concrete site of the confrontation, the scientization of time denounced by the burgomaster of Brussels does not appear to us as an insignificant confrontation: a technicoscientific program of the autonomization of time has clearly existed since the seventeenth century, and has just reached its planetary dimension. But to speak of this program as the putting into operation of a project that takes control of social time by way of scientific rationality would be to overshadow the social stakes with the same representation as that offered by the burgomaster of Brussels. This program is only comprehensible as participating in a more general process of an economic and political nature.[11]

The displacement carried out by the burgomaster of Brussels is thus in no way delirious; he is content to short-circuit the dialectic in operation, to oppose, as if they were terms existing in themselves, what are essentially two inter-dependent dimensions of the same development. It is a question of a typical case of what we will often encounter: a representation that is constituted by extraction, displacement, and naturalization.

Here one finds two terms—social time and scientific time—extracted from the common process whereby they are constituted, each of which is subjected to a particular mode of representation. Social time is rendered equal with a residual part of the natural and social phenomena that have actually contributed to its development and thereby produces an identification between nature and lived time. Scientific time, for its part, is separated from the social process that constitutes it and serves here to represent the process of the general commodification of activities, as if it was determined by a rationality that is external to the social and legitimated by science. The opposition between these two times itself functions as a way of splitting the process into two antagonistic metaphoric states.

The interplay of movements—extraction, displacement, and naturalization by way of the constitution of states—sheds light on the mechanisms at work in the enterprise of representation: never disinterested, they always refer to real processes whose stakes they simply legitimate, and ultimately obscure by making them appear as a state of fact whose evidence is rooted in a field that is external to this process.

However, to characterize as social the process that leads to the definition of scientific time obviously does not authorize a reduction of the technicoscientific to the social. Their two histories intersect, condition, and trigger each other, but each possesses its relative coherence and internal play of constraints.

From the Foliot to the Pendulum

From the point of view of the logic of technicoscientific development, we will show that the concrete object whose introduction marks the establishment of an autonomous law of time can be more precisely identified with the pendulum clock that Christiaan Huyghens constructed in 1658.

For the first time, a standard of time is constructed: the pendulum beats by the second. The second can become the elementary unit constitutive of all times, which from now on appear as simple multiples of it. The movement of the pendulum presents itself in clocks as the very law of time.

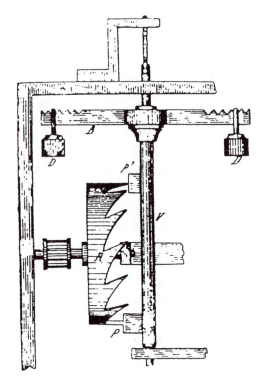

Fig. 2. Foliet clock. The regulators D are attached to the foliot
B. *Source:* C. Gross, *Escappements d'horloges et de montres*
(Paris: Dunod and Privat, 1913).

Previously, the mechanism that measured time, the foliot clock (see figure 2), appeared as a complex in which everything participated in the definition of the speed of the clock hands, without it being possible to specifically identify one element as the regulator.

We know that the problem resolved by the ensemble weight-escapement-foliot is to produce a continuous and uniform action in a world subjected to forces, thus a world of accelerated motion. The principle employed involves the breaking of the fall of a body into a succession of smaller falls that are incessantly stopped and started afresh: the movement of the weight is alternately freed and blocked by the escapement.

The foliot acts in this system like a counterweight, with its mass constituting a flywheel.

With each frontal collision between one of the foliot's two pallets and the escape wheel, the strike of the pallet against one of the wheel's teeth slows and blocks the wheel, thereby preventing the fall of the weight. The force of inertia of the foliot is exhausted by this operation, enabling the weight to resume its fall, driving the ensemble of parts and notably the escape wheel, which sends the foliot in the opposite direction.

The inertial motion of the foliot corresponds to a stocking of the energy communicated by the fall of the weight and will serve to block the fall of this same weight.

The jerking rhythm of the fall of the weight corresponds to the period of the alternate movement of the foliot, and determines the speed of displacement of the clock hands. As a result, their speed is a function of the totality of the mechanism, in particular of the mass of the weight and the inertial moment of the foliot. The regulators, modifying this inertial moment, allowed for the adjustment of the clock and particularly of the daily variation of the length of the "temporary hours."

It is usually claimed that Galileo's discovery of the law of pendular motion at last gave a scientific solution to the technical problem of the measurement of time. For oscillations of small amplitude, the period of the pendulum depends only on the length of the pendulum and the value of g, the gravitational constant.

Galileo's scientific definition of the law of pendular isochronism was not based on any rigorous definition (this would be produced by Huyghens), a fact that sheds light on the urgency of the demand for a timekeeping mechanism. Such a mechanism was necessary for the calculation of longitudes, which by the seventeenth century had become a major economic and political factor in the exploitation of the colonies, in the conquest of maritime space, and in the development of international commerce, all of which required the establishment of precise maps and the possibility of locating one's position with them.

However, Galileo did not produce such a mechanism: the free pendulum is a pure phenomenon; the oscillations need to be counted and the movement periodically restarted. The measurement of the pendulum's time of oscillation has no meaning other than in a scientific problematic, when it is a question of comparing the duration of precise phenomena. The pendulum is therefore not the timekeeping mechanism necessary for the calculation of longitudes. The pendulum clock is not a simple consequence of a scientific discovery.

Science and Technique

In 1657, Huyghens integrates the pendular phenomenon within a mechanism: the pendulum clock is born (see figure 3). Huyghens's work has been described as bring-

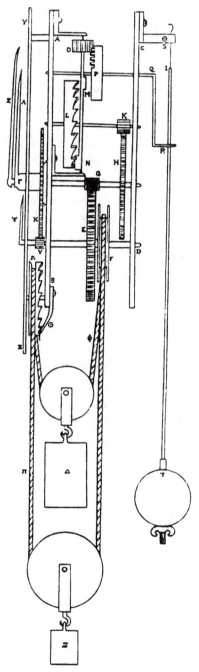

Fig. 3. Clock constructed by Huyghens. The pivot is connected
to the gears OP. *Source:* Christiaan Huyghens, *Œuvres complètes*
(The Hague: Nijhoff, 1988–1950).

ing about an encounter between a mechanism and an object, which thereby immediately embodies a law. This encounter instituted and began to realize a quite new
program: the subjection of a mechanism to a law that is external to it.

The foliot gives way to the pendulum, but it is not a matter of a
simple substitution: the foliot clock was part of a system, and it is as a system that it
determined time; the time of the foliot was only one element in the general negotiation that all the parties undertook between themselves.

On the other hand, the pendulum is the bearer of information;
one can even say that it is the information. Its autonomous movement will regulate
the rhythm of the escapement, which now has a double nature—both as the organ
that gives its impulse to the pendulum maintaining the law and as a receiver of
information.

The *Encyclopédie* of Diderot and d'Alembert clearly expresses what
will be from then on the program of clockmaking: "Nature having thus furnished
the means of measuring small parts of time with a nearly perfect exactitude, it is in-
cumbent upon the clockmaker's skill never to depart from this and to know how to
make use of it without troubling or altering the uniformity of these operations" (article titled "Horlogerie").

The work of clockmakers will largely consist of disconnecting,
as much as possible, the pendulum-regulator from the rest of the mechanism.

The most important part of this work will be concentrated on
the escape wheel, the only piece that must imperatively be in contact with the
pendulum.

As a result, the recoil escapement of Huyghens's era, in which
the pendulum is subjected to the action of the weight during the major part of its
course, is succeeded by the deadbeat escapement, where the pendulum is only subjected to the action of the weight at the moment of impulsion and of release, these
two moments now corresponding to one and the same interaction.

Next there appear free escapements and constant force escape-
ments. The free escapement brings about the practical realization of a project already imagined by Galileo: let the pendulum oscillate freely, beyond the two obligatory actions of release and impulse. The constant force escapement (see figure 4),
by stocking an always equal part of the energy produced by the fall of the weight,
renders the impulse independent of the action of the weight: the pendulum no longer
encounters the escapement mechanism, except to give it the release signal.

The technique has thus successively realized first an encounter
and then a disjunction between a law and a mechanism.

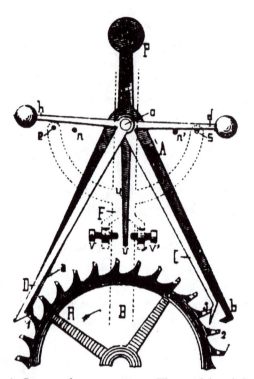

Fig. 4. Constant force escapement. The pendulum balance is indicated in dotted lines. The pendulum communicates information to the engagement wheel [*roue de rencontre*] through the intermediary of the adjusting screws V and V'. The energy is stocked through the intermediary of the arms C and D, alternately lifted and freed by the engagement wheel. The energy corresponds to the energy required to lift the weights h and d to a constant height of fall. The weights h and d, striking the springs e amd s, give the impulse to the pendulum. *Source:* Gross, *Escappements d'horloges et de montres.*

Galileo had only produced a phenomenon. Huyghens constructs a mechanism. But, as soon as constructed, the mechanical character of this mechanism is repudiated. The ensemble motor-cogwheels-escapement will only appear in the classical representation of the clock as the consequence of the imperfect character of the pendular device (friction, heating). The clock changes name, the part becomes the whole, the law replaces the function. From a finalized mechanism in which the rewinding of the weight was the price to pay in order that one could "tell the time," the clock becomes the pendulum, a mechanism tirelessly correcting the imperfections inherent in any local embodiment of the pure law of motion. It is to the extent that it cannot be abstracted from the world in which it is immersed that, in this representation, the pendulum needs a mechanism. From this comes the image of God the Clockmaker, and of the world he creates as an ideal pendulum that never needs rewinding. The metaphor of the World-as-Clock, so common in the seventeenth and eighteenth centuries, is thus based on the displacement that makes of a mechanism the realization of a law, and which, moreover, is indicated by the metonymy clock/pendulum.

Autonomization of Scientific Time

We have seen that this displacement in itself constitutes a technical program. When the clockmakers achieve the disjunction between the escapement and the pendulum, they develop a technical ensemble whereby the mechanism will actually become subject to the pendulum. But, on the other hand, the production of a lawful phenomenon, in this case the isochronal marking of time, has an ideological stake complementary to the one we have just described: the exclusion of the subject that produces technique in favor of the representation of an autonomous world in which human intervention no longer has any place. This exclusion will be redoubled with the exclusion of the subject that produces science, notably with the construction of that which an autonomous law implies, that is, a universal standard of time that transcends the particularity of relations between phenomena and enables them from then on to be expressed in an "objective," common language.

The law of isochronism appeared to offer the promise of a temporal standard guaranteed by a law. The oscillation of the pendulum had been presented by Galileo as univocally correlating the length of the pendulum and the period of the oscillation. Huyghens will discover that this is not at all the case: the law of isochronism is only approximate and only works for small oscillations. Huyghens's first attempts, the addition of "cycloids" to each side of the pendulum's point of suspension, aims to reestablish the rigorous lawfulness of isochronism by suppressing the influence of the angle of oscillation on the period. Note that this addition only modifies the pendulum's trajectory in a passive manner (see figure 5). It is a question of a purely geometrical assemblage that involves no mechanical contact (shocks, friction) because it does not imply any communication of motion.[12] Thus, it is a "correction" of a quite different order from that of the action of mechanisms that, according to the classical representation, must compensate for the imperfections brought about by the interaction of the pendulum with its environment. It constitutes the mathematical simplification of a lawful phenomenon, the suppression of one of the factors that determines it.

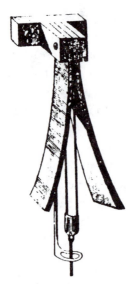

Fig. 5. Cycloids invented by Huyghens.
Source: Gross, *Escappements d'horloges et de montres.*

The cycloids clearly have no concrete application, and even less on a ship than elsewhere.[13] They nevertheless testify to the obsession exercised by the ambition to produce a universal law of time.

But, in this respect, the decisive step is certainly the redefinition of the relation between the cogwheels and the length of the pendulum.

The pendulum clock developed by Huyghens in 1657 had a period of oscillation of 0.743 seconds. This number has no raison d'être other than that it corresponds to a particular set of cogwheels. The length of the pendulum is entirely a function of the mechanism.

In 1658, Huyghens would invert the relation and now commence with a pendulum whose length is such that it beats by the second. The cogwheels are consequently defined as a function of the unit of measurement.

The measurement of time has in this way created a time that is autonomous in relation to the phenomena that are measured.

Previously, the strokes of the pendulum were able to be used to establish the proportions between the speeds of phenomena, or to calibrate the phenomena themselves. But the measurement that they provided had only a local value and necessitated the construction of an algorithm in order for it to have any sense elsewhere.

Once the second has been defined, one has at hand a reproducible unity constitutive of time and no longer simply a manner of relating phenomena. Time is no longer a measurement, it is the very norm of phenomena.

Objective, regular, normalized time, existing by and for itself, is born, uncoupled from what is now no more than the straightjacket of phenomena.

Work and Representation

Two histories have been recounted here. The one, sociotechnical, emphasized the social organization of *normalized* conduct; the other, technicoscientific, sought to show how a law gets the power to give its meaning and its reason to technique and how, by way of this law, an operation of measurement is transformed into an internal *norm* of phenomena.

These two histories, we have said, are inseparable. They both belong to the same historical process of putting human populations and natural processes to work.

Putting populations into (salaried) work as it has occurred during the modern era involves both a technicoeconomic infrastructure and a social formation. As any operation seeking to institute or reproduce a social formation calls for and engenders an ensemble of representations that will legitimate the relationships of forces that are produced, the process of "putting to work" work relies on such an imaginary: The peculiarity of this imaginary seems to us to be the legitimation of the idea of salaried work by way of concealing the no less real labor involved in assuring the conditions for a generalization of this salaried work.

We know that during the nineteenth century the equipment and devices necessary for the formation and reproduction of the worker were put into place. The school, military service, the morphology of the working habitat, and sanitary education functioned as matrices for the transformation of liberated masses of potential manpower into reliable and disciplined elements of the productive system. In particular, the metamorphosis of the worker's domestic space expressed the different strategies that regulated the relations between the production and reproduction of this system.

The scientific-technical arrangements that we have encountered—railroads, networks of electrical clocks, telegraphs, and time zones—are all so many devices that work at the articulation between representation and that which it conceals. On the one hand, just as with schools, habitat, and so on, they participate in the production of normalized conduct, but on the other hand, they appear to be simply the technical response adequate to preexisting technical and social requirements. It is in this double functioning that these apparatuses seem to us to embody what we have called the imaginary of "putting to work": inasmuch as they produce social conduct, they clearly participate in the movement toward the generalization of salaried work, whereas the representation that accompanies them presupposes this salaried work and thus denies the productive character of the constitution of this work to the benefit of a simple instrumentalist vision. Thus, the only labor that is recognized is that which necessitates the production of these "instruments" responding to what now becomes "social demand." And salaried work appears, as a result of the fact that it is not itself taken as the product of labor, as a condition, a natural human state. This operation of naturalization and generalization of salaried work, if it indeed leads to a measurement of human activities in terms of work, implies paradoxically that this work remains confined to the notion of the "human condition" and is strangely disconnected from anything that might be productive (*productive* should be understood in this text as producing effects other than the simple reproduction of the system).

It is in this manner that the imaginary of "putting to work" arrives at the metaphor of an ideal system represented as integrally put to work but not producing anything, that is to say, of a world functioning under the autonomous law of generalized circulation and exchange, of a world analogous, as we will see, to the World-as-Clock of classical thought.

But what invalidates this metaphor, the work of putting to work, the production of norms, of conducts, of canalization, of control, reappears analogously in the imaginary register in the form of fear and catastrophic predictions, fear of escapes, excesses, and of dissipation. We will return to this.

The metaphor of generalized and unproductive work obscures not only its distinctive conditions of functioning but also the most important dimension of capitalism: its expansive logic. This logic does not directly concern functions and applications: at its level, the utility of the steam engine or power hammer is not at issue; what is at issue is their power and the rhythm of their multiplication. The arithmetic symbols that put a number to this real process also function as a representation to the extent that they construct an image of expansion in a purely technicofinancial space and likewise conceal their own implication in the work of putting to work: the perpetual redefinition of new norms, new conducts, and new disciplines that they render necessary through the perpetual freeing up of the new social flows that they determine.

Here we will limit ourselves to the functioning of the metaphor of a world put to work under the sign of circulation. It will allow us to indicate the homogeneity of the tissue of representations that structure both certain socioadministrative discourses about the social world and certain scientific discourses about the physical world.

In both types of discourse will occur the articulation between two conceptual complexes: on the one hand, an ideal functioning conceived of as the pure enactment of the law that corresponds by definition to an optimal output; on the other hand, the divergence from the ideal, a divergence conceived likewise, a priori, as a source of loss, of waste, and of dissipation.

From now on, a primordial stake will be to know each time if the operation of putting to work bears on activities conceived of as essentially dissipative or, on the contrary, as fundamentally homogeneous to the problematic of putting to work.

From the Pendulum Clock to the Social Pendulum

We will begin with what, apparently, seems quite foreign to this operation of putting to work. The functioning of the clock as we have just outlined it corresponds to an articulation of the ideal and of the quite different dissipation from that which prevails in putting to work. Here it is a question of producing a mark on a clock face, and this mark can only be obtained through dissipation and shocks, whereby the energy of the falling weight is entirely consumed.

But it is characteristic that the dissipative collision that produces the mark has been repudiated to the benefit of a representation of the clock that is homogeneous to a world at work, in which all dissipation is imperfection. Huyghens and his successors, far from affirming the specificity of the clock, are on the contrary in agreement not to state it; for them, the pendulum must be kept for the sole reason that it is not perfect and not because its function of creating the marking of the hour excludes it forever and in principle from this ideal of perfection. From this comes the image of an infinitely skillful God the Timekeeper, whose clock knows no dissipation.

The World-as-Clock is a world in which everything works, in which the activity of each of its elements is conceived of as homogeneous to the law of work.

Thus one arrives at a paradoxical situation: in this world where everything circulates according to the law, where energies are exchanged without loss or deterioration, only one thing seems inconceivable: to pay the irreversible price of the mark of the law, to produce the memory of the law. And, as a corollary, it is just as inconceivable that the law would produce something. The world is an immediate embodiment of the law, with nothing left over, with no memory.

The world pendulum beats without anyone counting the beats.

But, at the Brussels town council, the clock has shown us that it also functioned as a "social regulator," and that this functioning included the putting into place of conducts and customs. In passing from the pendulum clock to the social clock, we pass from the representation of a world that is fundamentally homogeneous to the law, to the problem of an essentially restive world that requires disciplining.

One place where this problem will stand out strongly is the street: a place of noise, dissipation, collisions, and anarchic encounters and exchanges that it will be a question of transforming into a usefully controlled axis of circulation and exchanges.

Here, the law can be reduced to this injunction: "Circulate!" with, as a corollary, the phobia of all concentration, accumulation, excess, indeed, of any escape as the precursor of other excesses.

Material obstacles must be cleared away: "Obstacles, hindrances, ascents, and narrow crossroads quadruple distances and waste everyone's time without benefiting anyone. A straight, wide road with smoothly flowing traffic brings together and places in contact two points that appeared distant. . . . Those who spend millions can locate themselves only on an avenue that is clearly suitable for vehicles."[14]

Until then, the street, far from being a pure canalization, is a disparate site where debris and urban excrement accumulate; cluttered with stalls and broken up by steep alleys; where private and public space are mixed together; full of dead ends and hidden recesses. A dense life inhabits it: children, beggars, hawkers, sellers of countermarks, and ragmen all follow swirling trajectories transversal to its axis.

From now on, the injunction takes form and multiplies its effects. The plans for the alignment of houses follow; in Brussels, in 1846, the pavement regulations abolish cellar entrances that open from the pavement, basement windows encroaching on the public highway, doorsteps that overlap the pavement; "no sewer, no drainpipe, can overflow onto the pavement"; boundary stones and bowling are forbidden; doormats that jut out are prohibited.

The roadway and the pavement are constituted: linear, smooth spaces run between the clean facades of the houses demarcating them. The street begins to resemble a length of pipe.

Extracting the dissipative elements is the second logical time of this transformation. The Brussels markets will, on the one hand, be assembled in enclosures where exchange will be dense and regulated. The galleries, covered streets destined for commerce, cut straight through blocks of houses. Any possibility of earning some means of subsistence from the street will be eliminated. The Brussels rubbish bins are now only to be put out when the tipcart is passing. In Paris, the prefect Poubelle will implement the same regulation.

The street not only looks like a pipe, it functions like a pipe, like a duct.

The city, furnished with this circulatory system, can now claim to be an organism. The references, whether explicit or not, to Harvey proliferate during the course of the century. The law of blood circulation comes to naturalize the representation of a functioning town.

The representation inspired by Harvey seems to be a perfect example of an abusive, biologizing, physicalist representation. But we can note here the inadequacy of the critique of such representations when it is limited to questioning an unwarranted epistemological operation, or to denouncing the relationship of forces or of fascination between disciplines.

To speak here of the imperialism of the natural sciences is to not see the forest for the trees; above all, it is to not see that the metaphor of blood circulation accompanies and ratifies a very real labor whose object was the town. It is not the description of the town as an organism that should be the object of critical attention, but the sum of processes that enable this description to become more and more pertinent.

The law of blood circulation provides the program for an ideal town. The rectilinear movements of the circulation of men and merchandise are from now on uniform and regulated. Goods will glide without hitch from their source to their place of exchange; man will be animated by daily pendular movements:

Here one hardly ever sees excess; everything is modest, manners, customs, and entertainment. When, each morning, at the first light of dawn, these eight thousand workers leave the small towns or surrounding villages to come and take their place in the workshops, one does not hear any noise other than the road ringing under their feet. No shouting. No conversations. Everyone goes their way like people who have got nothing to say to each other and who dream only of arriving at their place of business. Their pace has the rhythm of marching troops: on return, when daylight falls or when the corvée workers return, it is the same movement. An exceptional country where the worker gives no other emotions to the entrepreneurs who employ him![15]

This description of workers marching toward the Krupp factories in the nineteenth century places in resonance the two senses of the word "conduct" on which our essay has played up to now.

Because, if it is a matter of canalizations, it is now clear that it is not enough to remove the obstacles such that a natural and harmonious movement ensues; what is still required is the production of actors in terms of conduct, habit, and behavior. The regulated pendular movement transports, without hitch or dissipation, the flow of the workforce from the domestic space to the factory, without emotion for the employers.

That being the case, one can understand that the great fear of the nineteenth century is the congestion and dissipation of the flows that it intends to control.

Concerning this subject, let's look at the account of a riot in Brussels given by the burgomaster of Brouckère in 1854:

> At that moment, I received notice that the superintendent of the twelfth division had been outflanked *by the crowd in the rue de l'Escalier, where, in spite of the efforts of the combined forces of the two first divisions, circulation had become impossible. . . . Finally, we had to deal with the rue des Pierres, where, in spite of all the warnings, the crowd continued to* gather. . . .
> *Two other officers intervened. One of them was greeted with chair legs, but finally the cabaret was cleared. . . . All the streets that converged on the square* overflowed *with people, there was* congestion *everywhere. . . . The gang recruited new members and ran flat out down the rue de la Pompe and the rue de Schaerbeek; they were met by the superintendent of the first division at the end of the rue du Marais and dispersed. . . . Overcome and chased, the gang, already well reduced in numbers at the junction of the rue de la Neuve, evaporated into the alleyways.*[16]

We said that the problematic of putting to work implied both the representation of a system in terms of a law and the concrete production of a mode of functioning that subtends and responds to this representation. From this point of view, the politics of the street, which implements the representation of the social system in terms of circulation, participates directly in the general problematic of putting to work.

Scientific Representations of "Putting to Work"

The representations attached to putting to work are not confined to the domain of human populations or economic processes. In the technicoscientific domain, the putting to work of natural processes can be considered as a guiding example that enables us to follow both the deployment of a metaphoric field and a set of arguments and formalizations, as well as practical realizations.

We have already said that the metaphor of the World-as-Clock was of a world wholly at work. In the world of dynamics, any effect of a force is, in identical manner, some form of work. In this case it is a question of "natural" work that does not require the condition of being "put" to work. Putting to work just means organizing naturally work-producing movements (that of weights or of water) at the service of human needs. In this sense, the *theoretical system* of dynamics constitutes the faithful and formal expression of the *metaphor* of work as a natural activity.

However, this must not allow us to forget that the concrete technical object "clock" cannot be reabsorbed into the dynamic model that assumes that the clock has an ideal reversibility. It is part of the very functioning of the clock to completely exhaust the potential energy of its weight through shocks and friction. Although clockmakers were able to significantly slow down the rhythm of this dissipation, they could in no way eliminate it. The intrinsic difference between the concrete functioning and the ideal model has, nevertheless, been constantly obscured.

During the nineteenth century, when it was no longer a question of putting to work masses in movement but heat, the relation between scientific ideal representation and technical object was both quite paradoxically conserved and severely destablized.

Heat does not naturally work. This is expressed in particular by Fourier's law, which describes a spontaneous process — heat propagates irreversibly; it dissipates without producing anything but a leveling of temperature differences. Fourier's law is thus the law of irremediable waste from the moment that the problem is to put heat to work in order to turn an engine.

The thermodynamics of the nineteenth century will be characterized by two irreducible elements. On the one hand, it will define and formalize the conditions of an *ideal* conversion of the energy of fire into mechanical work and the *ideal* output of such a conversion, but, on the other hand, it will be invested with the acute knowledge of the irreducible distance between the ideal that it constructs and the real functioning of heat engines. Extrapolating the consequences of this distance, physics will transform itself into the prophet of the irreversible evolution toward heat death, the state where no work is any longer possible.

Such a transformation is clearly not reducible to the simple awareness that a total economy of dissipation constitutes an inaccessible limit, leading to a theoretically well-defined absurdity (the zero productivity of the engine). Nevertheless, besides the "external causes" to which we will return, it is worth emphasizing the very real difference that separates the dynamic idealization from the thermodynamic idealization.

Whereas the ideal pendulum, which illustrates the conversion between kinetic and potential energies, may seem like the idealization of a natural object, the Carnot cycle, which illustrates the ideal conversion of thermal energy into mechanical energy, represents a completely paradoxical engine. In fact, any putting to work of heat presupposes a difference of temperature. But if two bodies with different temperatures are put into contact, it necessarily follows (Fourier's law) that the difference will be irreversibly canceled out: the heat dissipates without producing any utilizable effect. The ideal putting to work of heat thus implies that two bodies at different temperatures must never be put into contact. Now, the very principle of heat engines idealized by the Carnot cycle is that the mechanical energy is produced by the expansion of a heated milieu, following which the milieu must be returned to its initial volume by cooling (see figure 6). As a result, the Carnot cycle appears like a phantasmagoric representation: the engine system, during its isothermal phase, must give heat to a cold source or receive heat from a hot source while remaining at the same temperature as the source with which it interacts; it must modify itself in an infinitely progressive manner in order that at no moment will it leave thermal equilibrium. That is to say, it is a question of an engine whose perfection is quasi immobility. From the moment that this perfection is abandoned, the concrete conduct of an engine that approaches as much as possible the ideal, entails, at every instant, a set of extremely careful operations. And, however careful they may be, the result of these operations can only be dissipative. Here, as in the classical representation of the clock, irreversible dissipation marks the distance from the ideal, but it now also constitutes the very condition of a genuine production of work.

It should be noted, however, that the concrete functioning of the weighted clock allowed the problem of dissipation to be posed in similar terms: here too dissipation has a double status, being both the very condition of the functioning of the clock as a clock, in that its cogwheels must be periodically in contact with the "source" of potential energy (the weight), and since such a contact is always dissipative, a defect that has to be minimized by a technical program.

The status accorded to dissipation in the clock by the pendular idealization made it move entirely to the side of "defect." One can see here that another representation appears logically possible that illustrates not the trivial cycle of perpetual pendular motion, but rather, what we will call a *producer cycle* such as the one the Carnot cycle puts into theory. And, in fact, one can even say that the escapement clock is (at least in the Western world) the first mechanism to concretely

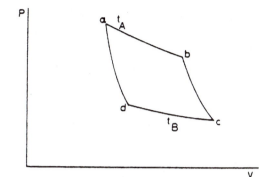

Fig. 6. Carnot cycle, functioning between tA and tB. Between a
and b, the isothermal phase, the system maintained at tA absorbs
heat and expands. Between b and c, the isolated system
continuing its expansion cools down to the temperature tB. These
two phases are "motor functions": the expansion of the system
can push a piston. Between c and d, the second isothermal phase,
the system is compressed and gives heat to the cold source,
at whose temperature it is maintained. Between d and a, the
system, once again isolated, is compressed until its temperature
returns to that of the hot source.

realize such a cycle; it is a device consisting of a periodic contact with a "source," the extraction of a certain quantity of energy from the source for the system, the transfer of this energy, the production of an effect by conversion of this energy, and the periodic return to the initial conditions.

Anachronisms

We are going to attempt here the deliberately anachronistic implementation of the notion of producer cycle in the description of clocks. The purpose of this is to bring out the fact that there is no simple natural relation between a technical object and its ideal scientific representation. In this case we will show that an ideal representation of a clock was conceivable starting from its concrete functioning, which illustrates perfectly the dimension of this functioning that is obscured by the pendular representation.

Let's take up first the description of the foliot device using the notion of a producer cycle.

The foliot is set in motion by the weight (extraction); the contact with the weight is interrupted, and the foliot now has only an inertial motion (ideally animated by a constant speed) whose kinetic energy represents the quantity

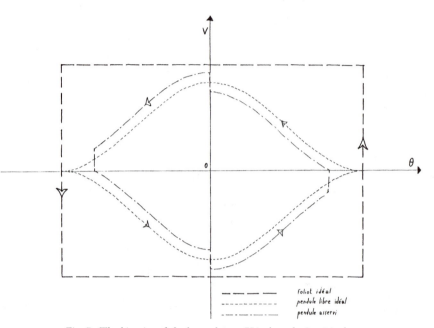

Fig. 7. The kinetics of clock regulators. V is the velocity, θ is the
angle characterizing the position of the regulator in relation to
the engagement wheel. θ = 0 corresponds to mid-course.

of energy extracted from the source; a new frontal contact is established with the
weight whose fall this time, instead of leading the foliot, blocks it. The energy of
the foliot has served to block the fall of the weight. At the moment when the energy
of the foliot is used up, the status of the contact with the escape wheel is reversed:
instead of being an obstacle, it becomes an organ of impulse.

The diagram (see figure 7) shows that, from the point of view of
the foliot, its inertial motion has "served" to cover the angle at the end of which it
will meet the obstacle that will absorb this inertial energy, cancel its speed, and allow it, by way of a new push in the opposite direction, to retravel the same angle
and rediscover its initial conditions.

The effect produced by this cycle is informational: the fall of
the weight is broken into segments and its jerks are transmitted by cogwheels inter-
posed with the clock hands that show the time.

One can therefore say that the energy consumed by the clock is
both entirely dissipated and converted into information.

The foliot clock draws the totality of its energy from the fall of
the weight, and, on each occasion, completely uses up this energy in returning to

the initial state. From this point of view, the substitution of the foliot by the pendulum is not the simple replacement of an irregular mechanism by another isochron,[17] but rather the putting into place of a device that, in braking and accelerating spontaneously, theoretically makes it possible to keep to a minimum the dissipative character of the encounters between bodies at different speeds. At each moment, the pendulum clock keeps a large part of its energy. In the final analysis, the additional energy provided by the fall of the weight serves only to compensate for the energy spent in producing the marking of time.

If one looks at the diagram in which the kinematics of the clock regulators is depicted, the situation that we have just described, that of a pendulum serving a clock mechanism, is represented by a hybrid curve in relation to the curves corresponding, respectively, to an ideal foliot (no friction, no recoil) and to a free and ideal pendulum. The hybrid character of the pendulum's curve expresses the compromise between information produced exclusively by the interaction between the parts of a system (foliot clock) and information carried by an autonomous phenomenon, but which needs an interaction to be realized.

We will reencounter this phenomenon in figure 8, which clarifies its irreducible character.

Figure 8 represents the energetic states of clock-regulating systems. The foliot, stripped of all potential energy, passes alternatively between a state whose kinetic energy corresponds to its inertial movement and a state of zero kinetic energy. The pure verticality of its back-and-forth movement in the diagram expresses the entirely constrained character of the energetic and informational variations of the foliot's movement.

The movement of the free pendulum corresponds to the diagonal expressing the conservation of the sum of kinetic and potential energies. The information contained in this movement remains wholly within it, without any possible exploitation. It is autistic information. In order for it to be extracted, it is indispensable that the diagonal be doubled and define a surface. This redoubling involves the addition of vertically composed segments to the extremities of the diagonal segments. These "vertical" segments signify that at the end of its course the pendulum gives energy by informing the system, and that at mid-course the pendulum is given a new impulse by the system.

What appeared as hybrid in figure 7 now takes on a dimension of universality: from the moment that it is a question of representing a cycle that produces effects, it is the cycles of the free pendulum and the foliot both reduced to one line that figure as particular limiting cases. The first case, that of the free pen-

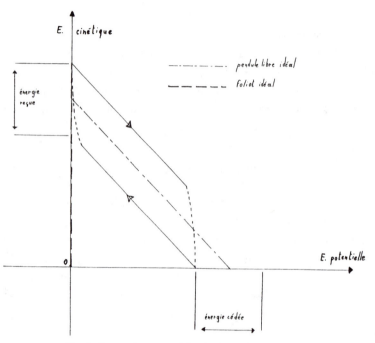

Fig. 8. Energetic states of the clock regulators.

dulum, becomes the limiting case of an informationally autonomous cycle that produces no effect, as no deduction of energy is effectuated. In the second case, that of the foliot, the production of effect involves the totality of the energy in each cycle. The general case appears as that in which a marginal quantity of energy is extracted and invested in the production of effect. Let us recall that this positioning of the problem that makes pendular motion appear as a limiting case of a producer cycle of effect has only been possible through the deliberate choice of a manner of reasoning that belongs to the nineteenth century as it illustrates the problem of producing an effect that has a *problematic* price, just like the conversion of heat into mechanical work.

We started with the problematic of dissipation and the question posed by the new status that it holds in thermodynamics. We began by remembering to what degree the phenomena involved in dynamics and thermodynamics are different. In the first case, it is a question of a motion in which the transformations of potential and kinetic energies occur in an essentially autonomous manner (the pendulum); in the second case (the Carnot cycle), it is a question of a series of transformations between thermal energy and mechanical energy that can only occur through manipulations, that is, from a series of essentially heteronomous transformations.

In the Carnot cycle, dissipation appears as a defect, the result of a less than perfect control of the system. But it also appears as a condition of functioning since perfect control leads to the absurdity of an engine functioning with infinite slowness, that is, a quasi-immobile engine.

We have now made explicit, at the level of an ideal representation of the clock, the same double dimension of dissipation both as defect and as condition of functioning. But, by the same stroke, a much more complicated logical operation has been produced. The pendulum idealization and the Carnot idealization both make irreversibility a defect but the corresponding reversible ideals occupy quite distinct logical positions.

By way of shocks and friction, the clock slowly but irreversibly converts the fall of the weight into a mark, just as, by way of a change of temperature and volume, the Carnot cycle converts heat into motion. In the clock cycle, irreversible dissipation apparently holds the logical place of the conversion idealized as reversible in the Carnot cycle, while there is no place in this reversible cycle for an analogy of the reversible motion of the free pendulum.

Information and Dissipation

Let's recommence and attempt to clarify this point. The Carnot cycle as a producer of mechanical energy from thermal energy enabled us to bring to light the lack of comprehension concerning the implications of the productive character of the cycle in the case of the clock. This enabled us to locate a profound analogy between the two cycles: in both cases, an essential stake is to minimize the differences at the moment of contact. The pendulum slows down by itself and is nearly immobile when it strikes against the escape wheel. Carnot's piston system is heated adiabatically prior to its contact with the hot source and cooled in the same way prior to its contact with the cold source.

Moreover, the productive dimension of the cycle, whether it is a question of the dissipative conversion of mechanical energy into information or of the reversible conversion of thermal energy into mechanical energy, appeared as associated to the *surface* of the cycle. We have seen the diagonal traced by the free pendulum spread out onto a surface from the moment it was a question of producing information from the fall of the weight; in the same way, one could imagine a Carnot cycle reduced to a line (one isothermal or one isothermal plus one adiabatical); in this case, the ideal cycle would no longer produce anything and would in this way be comparable to the movement of the free pendulum.

But the problem has just reappeared. Now it is the clock cycle that makes explicit the misappreciation by the Carnot representation of what is made possible by the thermodynamic cycle. If, even in its ideal representation, the production of information by the clock appeared as intrinsically dissipative, it is because it is the very definition of the production of information to leave an irreversible trace.

What illustrates the representation of the clock as a producer cycle is that, from the moment that there is a conversion energy/information (or the inverse conversion, which is produced, for example, when the pendulum strikes against the cog), there is a constraint brought to bear on a process, that is, in one way or another, there must occur a *putting into contact*, and any contact produced in this context must have a price.

Now, it is precisely a dissipative conversion of this type that implies the passage from a state of equilibrium to the immediately adjoining state of equilibrium, that is, the elementary link of the Carnot cycle. From now on, the Carnot cycle appears in its double dimension of energy and information, and its idealization as the smoothing of all the links where energy is converted into information and information into energy.

That the Carnot cycle can function in these two senses — producing work or transporting heat from a cold source to a hot source — in no way signifies a reversibility of energy/information relations: the reverse cycle does not "undo" the links of the engine cycle; it needs all of them for its conduct. In both senses, of its functioning, the links are intrinsically dissipative. In both senses, the elimination of the dissipation, that is, the absolute smoothing, implies an infinite number of links to be traveled, and, in this limit, the immobility of the system.

As for the clock, obviously nothing prevents one from imagining that an ideally careful conduct of the system can undo the mark and restore the weight to its initial height. But, in this case, the totality of the clock would function like an autistic pendulum and no information would be produced. Moreover, the necessity of a careful conduct of the operations signifies, as we have seen, the indefinite multiplication of the energy/information links. And, again, a paradoxical conclusion forces itself on us: the nondissipative clock is immobile.

Again we are playing with the anachronism, which enables us to conclude that the ideal representation of the heat engine that has prevailed was not the "natural" representation, but on the contrary denotes a strategy of occultation of the price paid for conduct and control.

In science we have twice encountered an operation of representation as a strategy: the clock as pendulum, the Carnot cycle as reversible.

We are not speaking of strategies in order to denote some conscious intention but to emphasize that this tension in representations does not seem to us to belong solely to the state of knowledge of those who produced them. It seems to us to be partly connected with the fact that representations and their objects involve a social practice.

Assemblages

From now on, the complexes object/representation/practice we will call "assemblages."

The location of a particular assemblage never constitutes an ultimate or complete explanation; assemblages are connected together in multiple ways and create communications between what is classically distinguished as different levels of explanations. Nevertheless, one can limit oneself, at a first attempt, to resubmerging the Carnot cycle and the pendular representation of the clock within a scientific-technical assemblage, that is, to taking them in connection with the concrete use of the device they represent: the one as representation of a heat engine, the other as that of an instrument of measurement.

The problem resolved by the Carnot cycle, to represent the functioning of a heat engine in such a way that the ideal output can be deduced from it, is not a trivial problem.

Prior to Carnot, the calculations of the theoretical output of a heat engine had been regularly refuted by the development of steam engines capable of performances that were largely superior to those predicted by the theory.[18] Carnot's aim is to raise the stakes to the highest point and conceive of an ideal representation such that any output greater than that predicted by this representation would lead to the absurd possibility of the "free" production of work (for Carnot, work produced without calorific displacement between two sources; for Clausius, work produced without this production being compensated by a flow of heat from the heat source to the cold source, that is, in contradiction with what will from now on be the founding postulate of thermodynamics: the necessity of two sources at different temperatures in order to turn an engine). In order to do this, Carnot will identify reversibility with ideality: it is to the extent that the ideal cycle is reversible that a hypothetical cycle of superior output would imply the possibility of the free production of work (by connecting this imaginary cycle to a reversed Carnot cycle).

The problem for Carnot is thus a problematic of engine output, that is, of the interrelation between the quantity of heat taken from the heat source and the quantity of mechanical energy produced. It is for this reason that Carnot cannot continue with a representation of a pendular type but creates instead the abstract notion of a producer cycle.

But as it is a question for him of output superior to all conceivable outputs, the reversibility of the system will become an a priori condition. And from now on the problems posed by the reversible passage from one state of equilibrium to another state of equilibrium will be obscured (the problem posed by what we have called the elementary energy-information link).

The problematic of output has never been foreign to the preoccupations of clockmakers: their *technical* program involves producing clocks that rarely need rewinding; thus the fall of the weight needs to be slowed down as much as possible. But this preoccupation is seen as purely technical, presenting no theoretical problem.

Furthermore, to the extent that the information produced by the clock is taken as pure knowledge that has no relation with the potential energy that it consumes, no ideal output that relates what is produced with what is consumed is theoretically deducible.

Moreover, as we have seen, the introduction of the pendulum will be accompanied by representations that obscure the significance of this problematic of output. The clock is no longer the producer of a mark but the display of an autonomous time. Seen from the perspective of a producer cycle, the pendulum represents a means of minimizing the quantity of energy put to work at each cycle. The pendular representation will obscure this effect and, on the contrary, identify the technical program of minimizing the energy necessary for the production of the mark, with a double program: to correct what is necessarily imperfect in the natural phenomenon and minimize the interactions necessary for this correction.

In this perspective, the whole problematic that brings into play the energetic dimensions of the functioning of the clock appears as entirely determined by the imperfect character of the pendulum.

We have just resituated the representations produced by the heat engine and the clock within what will be called microassemblages. In doing so, we have left the terrain of the internal history of the sciences. Representations of the clock do not have any meaning solely within the logic of the development of mechanical theories, no more so than the Carnot cycle in studies of the relations between heat and motion. Neither the "dynamic" object in itself nor the "thermal" object in itself appears to us to be entirely determining; both of them take on meaning within the stakes and specific problems actualized by their role in a technical problematic as instruments of measurement or as engines.

Of course, to emphasize that an object takes on meaning because it is part of a particular assemblage does not mean that it is created ex nihilo by this assemblage, nor that all the constraints and knowledges that constitute it are relative to this assemblage.

For instance, the mechanism of the foliot clock is not inscribed in a search for precision and regularity, but nevertheless, it is on the basis of this mechanism and the constraints and knowledges proper to it that Huyghens will be able to conceive of the first pendulum clock.

The study of a particular historical assemblage cannot disregard the technical, technological, and scientific lineages to which belong the objects that such an assemblage captured and redefined. Much more than an internal history, this study establishes a genealogy that locates the objects and knowledges belonging to other assemblages that will construct the new assemblage.

Thus, the Carnot cycle is not the normal outcome of theoretical studies on the output of engines (Poisson, Lazare Carnot), nor of those on heat (Black, Fourier), nor again of the empirical technical studies that led to the improvement of the effective outputs of steam engines (Watt, Smeaton).

Carnot *invents* a relationship that was not contained in any of these three lineages-disciplines and creates a new theoretical object, unexpected by any of them; unexpected for mechanics since one passes from the usual conversions between kinetic and potential energy to quite different energetic conversions, which imply a change in the state of matter; unexpected for the specialists of heat since its specific dimension of dissipation is systematically eliminated in the cycle that is represented as subjected to laws analogous to the reversible laws of dynamics; unexpected for the engineers who find themselves offered, as a model, an engine with zero productivity.

Neither can the study of a particular historical assemblage disregard the "macroassemblages" of which it is itself an element and which they too constrain and revive.

We have already encountered the determining importance of the economic and political interests involved in the conquest of maritime space for the constitution of a timekeeping mechanism, that of the establishment of networks of international exchanges (notably railroads) in the unification of time. It goes without saying that the extension of market relations since the Middle Ages is accompanied by the progressive institution of time as a measure of work, with all the techniques that this involves, from the clock whose hours can no longer be temporary up to the time clock.

We also know that the Carnot cycle is inscribed in an economic macroassemblage: thermodynamics is the daughter of the steam engine.

In both cases, the strategies of the microassemblages are marked notably by financial flows: prizes promised for a reliable method of calculating longitudes, the construction of observatories, prizes and subsidies offered for the study of the resistance of steam engines to high pressure, and so on.

The importance of the opportunities and demands stemming from macroassemblages is obvious, but we do not think that they are determining in themselves. They also need to be connected with representations that turn the macroassemblages' problems into problems seen as interesting and solvable. The possibility of a convergence between socioeconomic and technicoscientific interests is not a matter of chance encounters. When it occurs, it marks rather the existence of a common imaginary tissue that enables the appropriateness of ways of posing problems, positively identifies the types of solutions searched for, and engenders the ensemble of processes of naturalization, occultation, and metaphorization that we have encountered from the beginning of this essay.

We cannot prevent ourselves from establishing a connection between the fact that in the Carnot cycle the price of putting heat to work is silenced, the fact that information, although omnipresent, has the status of a nonproductive invisible source, and representations that presuppose salaried work and silence the fact that the operation of the submission of the social to the law of generalized exchange is in itself a labor.

One can make the hypothesis that the same coherence of the imaginary tissue is evident in the apocalyptic role that the themes of dissipation and heat death will play within physics after 1850.

Degradation and Utopia

Here again, an explanation in terms of a scientific-technical assemblage is possible.

On the one hand, for the majority of mechanical motor engines linked to preexisting natural fluxes (windmills, paddle wheels, etc.), the problem of reproduction does not present itself; the steam engine irreversibly and spectacularly consumes fuel that is known to exist in a finite stock.

On the other hand, the Carnot-Clausius cycle will bring to light the fact that the exploitation of thermal energy cannot happen without conditions:

there need to be two sources, one cold and one hot. In order for the cycle to be completed, any conversion of thermal energy into mechanical energy must be compensated for by a flow of heat toward the cold source. Now, the output, that is, the relation between the quantity of heat extracted from the heat source and the quantity of heat converted into mechanical energy, depends on the difference in temperature between the two sources. To the extent that the functioning—even the ideal functioning—of the cycle leads to the heating of the cold source, it appears to be diminishing the output and thereby "degrading" the machine. In ideal conditions, this degradation is reversible: the initial thermal difference can be restored by the cycle functioning in reverse. But the losses, the leaks, and the ensemble of dissipative processes that are inseparable from the functioning of a nonideal cycle also involve heat flows that diminish the thermal difference and degrade the conditions of functioning, this time irreversibly.

These two conceptually distinct phenomena (since the one belongs to the ideal representation and the other marks the divergence from the ideal), but which nevertheless produce the same effects, have been systematically and significantly confused. Carnot's ideal representation has been incorrectly viewed as signifying that heat cannot be exploited without irreversible degradation. After Carnot, many textbooks state that the cycles cannot be closed, that engines, even ideal ones, cannot autofeed themselves.

Such a confusion may be related to simple conceptual mistakes, the clarification of which led in fact to the definition of the famous entropy (which is conserved in an ideal Carnot cycle). However, entropy itself has usually been identified with degradation and loss. The fact that heat does not naturally work, that its work has a cost, be it idealized or not, is irresistibly connected with the nineteenth century's obsession with dissipation.

Here again, the explanation in technicoscientific terms seems to us to involve other assemblages.

Leaks, insulation, degradation—these three recurrent themes of the putting to work of heat in heat engines are identically the subjects of preoccupation that dominate another problematic of work, the putting to work of the populations of the industrial nineteenth century.

The fear of dissipation takes three dominant forms.

The oldest, without a doubt, is the fear of small leaks within the productive apparatus itself:

The objects of expense have so multiplied, that the least inaccuracy concerning each object would produce an immense fraud, which would not only absorb the profits but would lead to the loss of capital; . . . the least incompetence that goes unnoticed and that, as a result, is repeated daily, can become disastrous for the enterprise to the extent of ruining it in practically no time.[19]

But soon the apparatus itself must be isolated from its environment. It is advisable to separate what will become the working class from what will be nothing more than the dregs of society, the feared image of absolute dissipation:

How is it, that in spite of the considerable assistance given to them, the number of destitutes, far from diminishing, on the contrary increases incessantly? The answer . . . is easy: it is that up to now we have been exclusively concerned with caring for the poor, without hardly ever thinking of ways of preventing poverty. One could double or triple the amount of assistance, one could throw thousands into the bottomless pit of pauperism, but by persevering in this manner one would achieve hardly any improvement over the present state of affairs.[20]

Isolating the apparatus from its environment and making it function as a system involves the question of reproducing the workforce, and the fear of its degradation.

A significant research program was launched on this subject in Belgium in 1840 and 1843. In a report of the Brabant Medical Commission, one reads:

With a pale expression, sleeping in the open, often wasted away by misery and overwork, one sees these laborers, especially the children, prematurely aged. Their stomachs are swollen, they are pasty, their digestion is difficult; rickets and tuberculosis stamp them with the mark of physical degradation. Their chests are flat, their muscular system undeveloped, their intelligence nil; they are only interested in debauchery and depravation. The young women are tormented by verminous ailments; of pale complexion and sickly constitution, they are the victims of chlorotic and anemic diseases, their menstrual cycle is irregular, they are often incapable of becoming mothers, and, if they do become mothers, it is only in running the greatest risks for themselves and their children. Osteomalacia and rickets deform their pelvis; natural childbirth is often dangerous and sometimes impossible with the sole aid of the forces of nature.[21]

The ideal of functioning defined by Carnot for the heat engine seems to have for an analog in the social what we will call a circulatory utopia.

One of the dimensions of this utopia is the total incorporation of the worker into the system of production. In diverse and local forms, this dimension accompanies the history of capital. There will be the monastery-factories, the mining towns surrounded by walls, the model factories of New Lanark, and consorts where it is always a question of keeping the worker at hand and of responding to all his presumed needs in and by the factory. Here is a description of the Marquette mechanical weaving establishment:

A steam appliance for our workers' kitchen has been set up near the generator room.... The bakery provides for all the workers who want to get supplies for themselves and their families.... Bathrooms, supplied with hot water, have been put at the disposition of all the workers.... The children employed in the preparation of the weavers' weft are the object of particular care.... A school has been organized within the establishment.... The desire to keep our workers as near to us as possible, to shield them from the pernicious habits of the cabaret, to attach them more to us, made us decide to have a club built within the establishment itself that would bring them together. The ground floor of a building that contains the dormitory and the music room has been given over to a tavern/smoke den.... A billiard table, illuminated by the gas of the establishment, has been placed in the tavern. In the garden of the tavern, we have set up bowling, archery, and other games played in the country.[22]

But the logical extension of the same utopia appears like the putting to work, this time generalized, of an ideal à la Carnot: the complete integration of all activity within the system. This is a paradoxical extension because a completely regulated world, in which nothing would be lost, in which everything circulated, does not seem to be able to produce any other effect than its own indefinite reproduction as a system. This would then be, at the same time as its perfection, the death of capitalism. In any case, it is simply a matter of the phantasmic representation of one of the dimensions of capitalism, the ideal of closure, of a closed axiomatic, that in reality never ceases becoming undone, overrun, redefined by the perpetual mutation of flows that capital itself liberates.

In spite of everything, the circulatory utopia in all of its dimensions has locally acceded to a certain reality. Thus, in the Nazi concentration camps:

The life of work, reports a prisoner of Neuengamme, was not made up exclusively of insane acts. Productive work existed as such. But the sensible work was always slipping toward the insane. Purely mechanical work was not the alpha and omega of the prisoner's existence, it became in general the soul of his existence, just as, precisely, motion is the soul of any engagement of forces. That is why one also had to work in situations where there was no longer any work.... Not with a view to some product of work! But for the

continuity of the effort, for the motion itself. Not to be in motion signified sabotage.
We had to pick up rubbish, put it into piles, and then throw it around again
in order to pick it up anew.... "Because there must be motion!"[23]

In the social system reduced to the pure representation of the law of circulation, productive circulation annihilates itself in pure motion. The social pendulum beats indefinitely.

E   L   E   V   E   N

Drugs: Ethical Choice or Moral Consensus

(with Olivier Ralet)

Contrasts

THE SLOGAN "War on Drugs" is of course a quite paradoxical metaphor. The word "drug" designates a molecule that has effects if consumed, but that can have neither the projects nor the intentions of an enemy attributed to it. Who, then, is the enemy against which the war is declared? Is it a question of the possibility that human beings have always had, through the intermediary of drugs, but also through many other means, to produce self-transformations? Does it involve the consumer, the drug addict, the grower, or even the expert suspected of encouraging more widespread use? Or perhaps the dealer, the trafficker, or money-laundering banker? But then the paradox arises again: since these latter "enemies" only have the possibility of existing as a result of the state of war that has been declared ... The molecule only became a problem because of humans, and any connection between it and humans can be categorized as "mobilization against" or "collaboration with," but none of these categories "stands up" by itself, independently of the declaration of the state of war.

On the other hand, it is quite remarkable to find that the warlike metaphor could be applied with much less difficulty to the case of the AIDS epidemic. In this case, it is not a question of a molecule but of a virus, that is, of a living being, defined as such by "interests" that in this instance are at variance with those of humans, and by strategies that involve certain aspects of human lifestyles. This is why humans are quite rightfully unanimous in hoping for the vaccine or medicine that will signify the "defeat" of the enemy, or its eradication, as was the case with the smallpox virus. But one could have gone further. The men and women who, through their practices, make the AIDS epidemic possible could have been accused of favoring the strategy of the adversary and thus of collaborating with it. At the extreme limit, within the framework of a general mobilization, any human that refuses to take the recommended precautions could have been prosecuted for this reason. And one might have thought, considering the way in which drugs appear to be inscribed in the imaginary, following masturbation, as a corrupting curse on our youth, that AIDS would find a ready place: it would be the scourge sent by God, following that of syphilis, to punish the sinners and destroy Sodom and Gomorrah.

Of course, this kind of proposition has made the rounds and continues to do so here and there. But what might be striking is that, from the beginning, this proposition encountered an explicit and concerted opposition. It was, almost immediately, denounced both *as the trap that needed to be foreseen and as the trap that must not be fallen into.* Everything occurs as if our societies had managed to confront the problem of AIDS in a register that links the question of the epidemic with the ethical and political sense of the measures to be taken, and does this with an imperative: do not repeat history, avoid the repetition of exclusions and panics (the plague and syphilis), and thus identify what might lead to this repetition.

It is not a question here of pretending that everything is for the best in the best of all worlds, that AIDS patients are not the victims of any rejection, of ignoring the tensions and the permanent struggle that has to be kept up by the different associations representing the victims of AIDS. From our perspective, *the essential fact is that this struggle is generally welcomed as legitimate,* and that the problems posed by the epidemic are analyzed and dealt with, at least in Europe, on the basis of a clear and public perception of the problem: the threat of contamination will not be the alibi for coercive measures (obligatory testing, isolation units) that would reassure the "public" but would constitute a historical regression for our societies. This is what we will call the AIDS *event.* By event, we designate the political and ethical *choice* affirmed by the refusal to allow oneself to be forced by the factual existence of a virus to take a step backward and, for example, designate homosexuals, blacks, and drug addicts as "others."

First and foremost, to combine the notions of event and choice implies that no instance—whether political, ethical, of the mass media, or technical—can be said to be the "author" of this choice. Because in this case, it is much rather the event itself that has decided the manner in which these instances would be articulated. Many accounts enable one to follow the history that has led to this choice, its hesitations, and the relationships of forces involved in them. No account can have the status of explanation, conferring a logically deducible character to the event, without falling into the classic trap of giving to the reasons that one discovers a posteriori the power of making it occur, when, in other circumstances, they would have had no such power—what Bergson called "the retrograde movement of the true."

The "AIDS event" is thus inscribed in the ethical and political register. Should one now think that, since ethics has prevailed, technique and the experts who are its spokespersons have had to stand down and take second place? Although those who place ethics and technique in opposition might think so, this has not happened at all, quite the contrary. The decision to try to not repeat history leads to the question of *how* not to repeat it, which is a strictly technical question.

The "AIDS event" is characterized by the choice of not yielding to the urgency of the strictly medical problem, of resisting demagogic and security-seeking temptations, in other words of trying to actually *pose the problem clearly*. This is why it has been decided to give a hearing not only to those whose expertise represents the virus and its paths of transmission but also to those who represent what we know about the manner in which individuals, groups, and societies invent themselves by way of rules, laws, and technical apparatuses.

By way of illustration, we will quote Antoine Lazarus, professor of "public health" medicine at the Bobigny faculty:

The prevention of AIDS in its present phase suggests an original strategy. Faced with an infectious danger or one of diffusion by way of example and use, as is the case with the use of toxins, our country has sought to propose (before AIDS) preventive measures implemented by the community: obligatory testing, the isolation of infectious subjects, repression of the incitement, vaccination campaigns, and so on. Now, for AIDS the key idea that is essential for prevention is that of individual prevention. Following the advice of both French and foreign experts, as well as the advice of the National Ethics Committee for Research and Biological Experimentation, the prime minister, Michel Rocard, repeated this in his televised speech of December 8, 1988.

Following his explanation, it appears that the imposition of obligatory testing would not give good results, neither at the technical level, because not everyone would submit to it, and furthermore it would be necessary to continually repeat the obligatory examinations that we know, even when negative, not to be the immediate proof of noncontagion, nor on the psychological level, because the population would believe itself to be protected by effective general

measures when, in fact, this would not be so. That means that today, in this particular case
(but it has strong value as a model), the most efficient solution for both individual and
collective prevention is information campaigns about measures that are easy to access,
but freely chosen. Let everyone do what is necessary, let everyone take protections in his
or her own way. The risk is everyone's; protective measures need to be taken
by everyone. No prohibitions or constraints are capable of giving good results.
The instrument of prevention, considering the diversity of characters, of situations,
of cultures, of ambivalent temptations of life, and of playing with death,
is an individual responsibility.[1]

That, in this case, the assembled experts came to privilege individual responsibility does not mean that they held as an intangible premise the ethical imperative of "respect for individual freedom." What they took into account were the "psychological" consequences, for the citizens, of a legal apparatus that would make them believe that they were protected but that did not require their responsibility, and they did so *because those who were interested in such an approach were present among them and had the same right to a hearing.*

Thus, the "AIDS event" is not characterized by the submission of technique to an ethical choice, but by the type of *technical* definition brought to bear on the problem, which is also to say, by the type of experts recognized as legitimate spokespersons in the discussion. The idea that where ethics prevails, technique must submit, is consequently replaced by a quite different relation. Technique does not have to submit because it is never in control. The one who controls is the one who determines how the technical problem will be posed and notably if and how it will take into account constraints determined by human values and interests. This determination of the problem is a question of political choice.

The classical opposition between ethics and technique is based on the idea that a "purely technical" solution would be, by nature, foreign to the "truly human," would treat human beings like things, without regard for the choices and values that by nature elude its calculations. But, when one describes a solution as "purely technical," the question always arises: what has been purified and in the name of what? In the name of what political choice, for example, has "economic rationality" authorized the disregard of the problems posed to inhabitants and to the environment? One need not oppose to a "purely technical" solution values that would transcend technique; it is always possible to maintain that this "purely technical solution" is a solution to a problem that is *technically badly formulated,* that is, to a problem posed according to certain a priori imperatives that have resulted in hand-ing over control to certain experts and in ignoring others. To those experts who are ignored correspond the dimensions of the problem, judged as irrational, illegitimate, or immoral, which it has been decided not to consider. And to those whose interests are, in this way, judged as irrational, illegitimate, and immoral, the power authorized by expertise is not addressed as it would be to citizens but solely as it would be to potential offenders.

Nevertheless, this does not mean that every interest that is taken into account will be satisfied. In the case of the decision bearing on the mandatory wearing of seat belts, when the experts decided, rightly or wrongly, in favor of the need to restrict individual freedom, that is, to subject the authorization to drive a car to certain conditions they placed some interests ahead of other clearly legitimate ones. The main thing, from our point of view, is that this decision created—and, moreover, still creates—controversy. In other words, the reasons for the decision were able to be communicated to the interested parties without violating their dignity and without denying the legitimacy of their objections.

What about the case of drugs and drug addiction? In this case also, experts were brought together, but with quite different results. As Marie Andrée Bertrand reminds us:

At the end of the 1960s or the beginning of the 1970s, no fewer than a dozen countries had proceeded, through the intermediary of national committees or commissions of inquiry, to an examination of what was from then on called "the drug problem": its extent, its causes, and the means of remedying the situation. None of these committees or commissions, except perhaps the Pelletier Committee in France, had recommended maintaining the status quo to their respective parliaments. Some argued for the decriminalization of certain substances, others for the abolition of the crime of simple possession, and so on.

Nevertheless, nowhere, in any country, did the reports of these committees have an important effect on the legislation. It is true that the actual application of the law was modified in some places and that there was a certain de facto depenalization, but with all the arbitrariness that that implies, that is, the continuation of legal proceedings when the user or small-time dealer is a foreigner, marginalized, or quite simply when the policeman or judge did not like their appearance.[2]

Thus, in a general way, one can conclude that, in the case of drugs as with AIDS, the experts were assembled, but that in the first case, the type of history that enabled Antoine Lazarus to quote Michel Rocard, himself quoting expert opinion, did not take place.

This does not mean that the laws that defined the official policy on drugs in the various countries had ignored the technical arguments. Take the case of the law of December 31, 1970, which defined and still defines French policy on the question of drugs coming from the South. This law, concerning the health measures of the war against drug addiction and the repression of the trafficking and illegal use of harmful substances, was adopted unanimously. A unanimous vote clearly expresses a problematic of mobilization, especially when one remembers that it was actually a question of a "law of exception."[3] The explanation for the reasons behind this law has an entirely "technical allure." In particular, it is a technical argument that allows for the justification of the penalization of drug possession for personal use. Such a justification was necessary, remarked the Pelletier Report in 1978, because "common opinion" might be surprised that acts that only involved the person indulging in them should be subject to repression. Let us observe, as does Caballero,[4] that what is designated quite condescendingly as "common opinion" does no more than take up the principle stated in article 4 of the Universal Declaration on Human Rights, according to which "freedom consists in being able to do anything that does not harm others." The argument invoked in support of the law is the social cost: "in a period when the individual's right to health and health care is increasingly recognized, in particular with the generalization of Social Security, it is normal that, in return, society can impose certain limits on the way people use their own bodies, especially when it is a question of substances whose extreme harmfulness has been unanimously denounced by specialists."

However, if the social costs are invoked, the technical question of these costs and the different methods of managing them is not raised. The experts were not asked about the different ways of reducing them. They were not asked if legal prohibition was the best way. They were only asked to testify to the "extreme harmfulness" of the incriminated substances, and, we are told, the specialists were unanimous on this subject. Regarding AIDS, doctors were assembled, but also historians, sociologists, epidemiologists, and psychologists — in short, all those who represent the social body as it is affected by the problem and as it might be affected by the solutions that were to be proposed. In the case of drugs, the opinion recognized by the law is purely medical.

For all that, the "extreme harmfulness" recognized unanimously — where did they find specialists capable of such touching unanimity? — is less a technical argument than a useful justification. If not, how can one explain the fact that alcohol and tobacco are legal but not cannabis? Moreover, in 1978, the Pelletier Report did not hesitate, concerning the harmfulness of the by-products of cannabis, to reverse the burden of proof, that is, to recognize that, in their case, it is not their "extreme harmfulness" that the specialists can testify to, but the impossibility of demonstrating their complete harmlessness: "As for the innocuousness of the by-products of cannabis, it is not because their dangers are not clearly understood that it should be taken as established." It is true that, as we will see, the Pelletier Report, whose singularity Marie Andrée Bertrand emphasized in that it was the sole report to argue for the need for a legal status quo, had in the meantime put into place a new type of argument concerning this subject. A new type of expertise had been recognized as legitimate that made it possible to justify the prohibition of cannabis by-products by in fact presupposing their innocuousness.

Quite clearly, the law of December 31 is not purely repressive, but it plays its part in the technical question of the treatment of those that must be considered as much victims as offenders. It institutes, notably, the therapeutic injunction, a measure that gives the public prosecutor the possibility of offering a "choice" to first offenders convicted of drug use: a choice between prison and a detoxification cure accompanied by medical supervision and psychotherapy. There are very few psychotherapists in the world who would be prepared to attribute much chance of success to a psychological treatment practiced under constraint (here to avoid prison), for anyone in general and in particular for a drug addict. The ineffectiveness of the measure was thus predictable. That it is a question of a measure with a technical *allure* has, moreover, become a quasi-explicit fact. If not, how can one understand that some public prosecutor's departments—such as Lyons—implement the therapeutic injunction with regard to smokers of hashish, who are not generally recognized as "drug addicts"? The main interest of the measure is in getting the illicit drug users themselves to settle the false old debate (or the real word game) with which it has been decided to define their case: "are they delinquent or ill?"

Thus, drug legislation clearly expresses the submission of technique to politics. More precisely, the political choice of experts recognized as legitimate allowed the term "expert judgment," authorizing a law, to be applied to a purely medical statement concerning the "harmfulness" of the incriminated substances. Once the experts have made it possible to justify the statement "you must not," legal and technical measures could also submit to this premise. Now it will be said of these measures that "they must work," even if many people think that they will not. And any public controversy on this subject will expose those involved to the accusation of "demobilization" or of "collaboration with the enemy."

AIDS/drugs: on the one side, a debate and a noncoercive choice, and, on the other side, a consensus about an obvious coercion. Are there, in this case, "two" politico-ethical choices, one against drugs, the other against anti-AIDS coercion, two choices commanding two different relations with expert opinion, and which we would have to do no more than take note of? There have clearly been two "events," but this would only lead to a symmetry of judgment if the choice of subjecting expert opinion to the imperative "don't take drugs" was to be accepted as an ethical choice.

"Don't take drugs": when a statement like this is made by a legislator, is it a matter of ethics? We maintain that it has nothing to do with ethics. If there is a difference between ethics and morals, it is clearly that morality is concerned with statements like "must one," or "must one not," whereas ethics must, above all else, ask the question, "Who am I to say to the other 'you must,' or 'you must not,' and how will this statement define my relation to this other?" In this instance, the "I" who must, ethically, ask this question, is none other than the legislator, the one who states what the power of the law will be over citizens. This is why ethics, in political matters, is judged less by the types of solutions that are proposed for problems than by the way in which the positioning of the problem and the solutions proposed situate and involve those to whom they are addressed.

Thus, the question of ethics when it comes to the management of the community bears above all on the choice of a type of society. There have existed, and still exist, societies in which individual ethical choices are supposed to be inferred from the collective norm. It happens that our societies bear the challenge of a great refusal, that of identifying the collective instance with a normative instance. This is the very choice that defines the requirements and the risks of democracy and it has nothing to do with the institution of a relation of indifference between the community and the individual, that is to say, the hyperliberalism of "destroy yourself in complete freedom from the moment that it does not harm other people." The singularity, with respect to the management of the community, of the democratic choice in relation to other possible choices, lies less with the question of content than with the question of modality. If this choice does not in itself dictate any "solution," it does, on the other hand, demand that, in the definition of solutions, that is, the methods of community management, one fact cannot be ignored: the fact that each proposed solution anticipates and suggests, through the way it addresses the individual, what this individual is and what he or she can do.

Let's take two French examples of this anticipation-suggestion (one in each of the registers that we are examining: drugs/AIDS). Dr. Francis Curtet, a French specialist in drugs, criticized his former employer, Dr. Olievenstein, for having written that there are heroin users who control and manage their own consumption of the drug. When Dr. Jean-Paul Escande dared to announce that the news on the AIDS front was not that bad, he was treated as irresponsible, without it being possible to determine if this "irresponsibility" came from the false nature of the information or from correct information that needed to be kept from ordinary people. In both cases, the criticism expresses the fact that acceptable statements are those that anticipate-suggest the stupidity of individual choices: if one tells people the truth, they will not be "reasonable"; thus, they have to be lied to in order that they will behave reasonably.

Obviously, the temptation is strong, when the majority of citizens stick to a conviction (for example, "don't take drugs"), to consider that it is good democracy to give this conviction the force of law. But the democratic ethic, which, in our society, is the only source of legitimacy for the political process, is not a matter of content or of a pure and simple submission to majority rule. It is the manner in which the political process addresses citizens, that is, anticipates-suggests what they are and what they are capable of, that will determine whether the democratic ethic is denied or affirmed. In other words, a constraint follows from the democratic choice to which our societies claim to owe their definition: any method of management that implies the supposition-anticipation-suggestion of stupidity or infantilism of the individuals that constitute this society should be excluded; for, if they are defined as stupid or infantile, democracy itself can only be defined as manipulation, a modern, new way of leading the flock.

The direct consequence of this constraint, which for us defines ethics as far as politics is concerned, is nothing less than the need for a truly technical positioning of the problems of community management. It should be understood that by "truly technical" we do not mean the positioning of the problem by a group of experts, who might define themselves or might have been defined as legitimate authorities. When it is a question of the complex problems posed by society, the only reason why some experts may claim "authority" is their alliance with power, that is, when power has determined the manner in which the problem "should" be posed. On the other hand, controversy between experts, which is the only truly technical way of defining a problem, engenders not only precise information about the "reality" and its parameters. It imposes not only a process of explanation of its many dimensions and a setting forth of the risks of the possible choices, their presuppositions, and their consequences for reality. It gives rise not only to the invention of aims and means whose pertinence can be discussed and evaluated. It anticipates and suggests that citizens have the right to expect and demand that their diverging interests be taken into account in the controversy, but also that they are capable of understanding that there actually is a controversy, capable of being interested in it and not getting into a panic or becoming demoralized when confronted with the "reality" of the problem. The technical controversy, by nature both public and publishable, creates the conditions presupposed by the exercise of democracy.

If the ban on drugs in France was not the product of a technical positioning of the problem of drug consumption and addiction, it is primarily because it was a question of giving the force of law to a *moral consensus*: "don't take drugs." Moral consensus is always expressed in terms of general slogan-phrases, which designate the one who would not be in agreement as a public enemy or, and notably if it concerns a dissident expert, as irresponsible, and which blur the distinction between community management and individual choices. It defines the "evidence" that must be shared by everyone and not a choice, open to controversy and discussion, formulated in such a way that all those concerned by this choice are addressed as both legitimate parties to the problem and capable of being interested in the process of solution.

To the primacy of the evidence on which the moral consensus is based and maintained there corresponds, at the same time and indissociably, the neglecting of the risks and ethical demands of democracy in favor of a logic of mobilization, which is, of course, one of the forms of the art of leading a flock, as well as the a priori distrust of those who are interested in the "reality" that is always capable of inspiring doubts and complicating the relations of adhesion and belief, in short, of demobilizing the suggested simple consensual faith.

When this primacy of the evidence is carried to its extreme point, it designates what is called the logic of the "Freudian cauldron": the nature of the argument is irrelevant as long as it leads to the desired conclusion (the reader may remember the anecdote cited by Freud: someone complains that the cauldron he lent has been returned with a hole in it. The borrower replies, more or less in this way: "First of all, I've returned your cauldron intact; second, this cauldron already had a hole in it when you lent it to me; and, finally, you never lent me this cauldron").

It so happens that the Pelletier Report, which, as one will remember, was cited by Marie Andrée Bertrand as the only expert report that concluded with the need to maintain the legal status quo concerning drugs, provides, in its discussion of the proposals for the depenalization of the use of cannabis by-products, a good example of this "logic of the cauldron": (1) Drugs are prohibited because they are dangerous and, "as for the innocuousness of the by-products of cannabis, it is not because their dangers are not clearly understood that it should be taken as established" (p. 207); (2) If the use of cannabis expresses a protest against the dominant order, "is there any interest in withdrawing its proscribed character since we are dealing with a desire for transgression that will inevitably be turned into other more harmful behaviors?" (ibid.) ("more harmful?" Is the innocuousness of cannabis taken as established here?); (3) "For all that, the controversy that has taken place around hashish is to a large extent a false problem, since in reality the use of 'H' is not as severely repressed as some people claim" (p. 208). Repeated in different words, this demonstration gives the following: first of all, cannabis must be prohibited because it has not been proven that it is not dangerous; next, cannabis must be prohibited because it is not dangerous and our youth must be left the possibility of transgressions without danger (in the absence of which they will transgress dangerously); moreover, cannabis is not "as prohibited" as is claimed (see, on this subject, the conclusion of Marie Andrée Bertrand's citation).

We will return to the highly significant point number 2. First, we will show that the "logic of the cauldron" is in no way an uncommon digression but is well anchored in what can be called the French ideology of drugs. In an interview with the *Revue d'Action Sociale* (January–February 1985), Dr. Francis Curtet (already cited) took up the debate.

REVUE D'ACTION SOCIALE:
Why do you think it's preferable that the sale of hashish should not be legalized?

DR. CURTET:
Because it's a product that in spite of everything is dangerous, obviously not as dangerous as heroin, but nevertheless dangerous. There are four dangers:

1. in high doses: memory problems, lung pathologies (as with tobacco);

2. the risk of imprisonment and prison, which is much more toxic than hashish itself;

3. the danger of being solicited for other products the day the dealer says that there's no hashish;

4. the most worrying danger: hashish acts like a "disconnecter" in relation to reality and, for young adolescents in the process of constructing their lives on every level, the risk is great of wanting to "get high" rather than deal with difficulties (educational, professional, sexual . . .).

Argument 4 remains valid if one replaces the word "hashish" with the word "television" and "get high" with "watch television." Thus, arguments 1 and 4 justify the prohibition of hashish by comparing its dangers with those of perfectly legal drugs: tobacco and television. One could obviously maintain that if it is in fact difficult to prohibit tobacco and TV, it is precisely a question of not adding other bad habits to these. But in that case, arguments 2 and 3 go against the contention: hashish is more dangerous than TV or tobacco precisely because it is prohibited.

Another recurrent theme in Dr. Curtet's statements and publications is the importance for adults of not being discredited in the eyes of young people with responses that "miss the point." We applaud this theme, but fear that it can only be understood as one more example of the "cauldronesque" succession of arguments that have been employed. If anything is being suggested to Dr. Curtet's readers, it is clearly that it is expected that they accept that the law of 1970 is not open to debate and that all argumentative approaches are of equal merit from the moment they lead to the affirmation of the moral consensus. In such a cauldronesque logic, moral consensus and the art of leading the flock converge: the fact that the logically vicious character of the argument does not even try to conceal itself presupposes-anticipates-suggests that the drug does not constitute a problem that could be discussed, but the object of a conviction that must be shared.

Now let's return to argument 2 of the Pelletier Report: the prohibition of the use of cannabis has the advantage of offering the possibility of satisfying a desire for transgression in a relatively harmless way. This argument, presented as "technical," is in fact blind and deaf to what we have learned through the Dutch experience: young people, deprived of their "transgressive lure" through the depenalization of the consumption of "H" do not turn to more dangerous drugs. But of course this argument was not produced by the experts in the field, nor deduced from the accounts of the Dutch experts. It expressed the fact that, from then on, new types of experts were recognized as the source of authority on drugs by the political power: psychologists and psychiatrists for whom taking drugs, whatever the drug, is not harmless, one has to attend to it—that is, as the commission well understood, maintain the prohibition.

This controversy [on the false notion of "soft drugs"] has ignored the fact that a consideration of the user's behavior is much more important than that of the product used. However, it is impossible to reduce the consumption of hashish to an ordinary act, stripped of deeper meaning. From the moment it becomes habitual, it reveals personal difficulties and cannot be explained by the simple quest for a fleeting pleasure. (P. 208)

Political changes in France have hardly affected the way the problem of drugs is presented because, in the Trautmann Report addressed to the Socialist prime minister, Michel Rocard, the same type of argument was used quite recently against antiprohibitionist proposals:

The person who takes drugs is expressing something and giving out a signal (identity problems, anxiety, difficulties in dealing with life). This implies that one listens to the message, that one cares about the person, and that one takes the time to help him or her. Why, in these conditions, do the abolitionists refuse to listen? Why do they pretend to think that one is only faced with a problem of behavior? And this at a time when the considered opinion of those who are really in contact with drug addiction aims not only to abandon the approach of categorization (legal-illegal) but also to resituate drug addiction within the ensemble of our youth's expressions of anxiety. (P. 201)

The only difference between the two texts is perhaps that the Pelletier Report had been innovative in appealing to new types of experts, whereas, at the time of the Trautmann Report, these new "psy" experts had acquired in France the position of authority on the matter. Obviously, there has been progress in the "destigmatization" of the drug addict that these experts have brought about: he or she is no longer a monster but a case. But the progress equally concerns the justifications for the law of 1970. From now on, the argument about the "extreme harmfulness" of the products does not need to be "forgotten" when it comes to the prohibition of cannabis by-products. The moral consensus of "don't take drugs" that the law expresses has found a new and original justification: the harmfulness of the product is not what counts primarily, but the "call for help" that, quasi-ontologically, defines the "drug consumer."

As we have said, the acceptance of the risks and requirements of democracy is verified through the acceptance of a technical positioning of the problem to be resolved, whereas the moral consensus blurs the distinction between community management and individual choice, defines the community as a flock to be led for its greater good, and chooses experts who are likely to confirm the justification of this good. Whoever reads the expert opinions mentioned by the Pelletier Report as well as the Trautmann Report would think that their authors were pleading for an increase in the budget allocated to their specialty and to the centers dealing with prevention or assistance to drug addicts. The idea that "taking drugs" can be a sign of anxiety that should be listened to, and the "double wish" affirmed by the Pelletier Report "to avoid the 'banalization' of the use of 'H' and its 'dramatization'" (p. 208), clearly constitute objectives that a psychologist might propose to parents worried about the difficult phase their kid is going through. But this double wish appears here in a report addressed to the state, and it is not related to the demand for subsidies destined to help those who ask for such assistance, whether their problem involves legal or illegal drugs, but to the affirmation of the legitimacy of the law that categorizes drugs.[5] When the state chooses its experts from among those whose job it is to delve into hearts and heal souls, when it chooses to listen to them in a register that suits it, and when it starts to play at being psychologist itself, the die is cast. At this point, quite strange conjunctions are celebrated: "A consensus exists today, notably between judges and doctors, on the notion of the law as both a point of reference and a reminder of the reality principle."[6] In this way, the state puts itself in the position of judging what secretly animates those who violate the prohibition, transgressing the law that it has enacted, of understanding, beyond particular acts, the "message" for which it now becomes the addressee. From then on, words such as "law," "transgression," or "prohibited" acquire an irresistible double meaning. At the end of a fairly breathtaking trajectory from the psychoanalyst's office to the public prosecutor's department where the "cases" are deferred, those who reject the consensus designate themselves as needing to be taken charge of, like "cases," appealing to the state to take up the former responsibilities of the king, "father" of the nation, or of the church, the benevolent and firm shepherd guiding the faithful flock.

"The law, a reference point structuring young people in difficulty: young people in difficulty often have in common parents who have not helped them to integrate the law in a profound way. According to psychiatrists the prohibition has a structuring effect on our youth, who are clearly weakly structured."[7] It is not immaterial that the question of the protection of young people is systematically confused, here as in the other reports, with that of laws suppressing the use of the drug for all age groups.[8]

As was again maintained by Michèle Barzach, the psychoanalyst minister of health for Jacques Chirac's government, those who contest the law of 1970 confuse "limits" with "prison cages."[9] The state has fulfilled its paternal role in setting limits whose first meaning is, as one knows, to structure the subjective life of the "child." That the child claims that these limits are arbitrary in order to accuse them of being obstacles to his or her freedom is enough to show that he or she has a problem to which the state cannot remain indifferent.

It is rather uncommon for experts to heroically refuse the bait that power holds out to them, the designation of their knowledge as being the only pertinent one. Consequently, we have seen flourish a literature of psychoanalytic appearance, typically French, with Lacanian references, whose psychotherapist authors choose to forget that their field of competence is limited to those who have asked them for assistance and bravely assimilate drug taking with revolt against the "law of the Father," or infraction against the juridical law with transgression of the symbolic law[10] in such a way that the downhill slope from drug use to drug addiction appears direct and fatal.

There is no doubt that during the course of a psychoanalytic cure, it can seem that the law violated by a drug user is not, in his fantasy, one law among others but the Law; what any consistent psychoanalyst also knows is that an elucidation of this type only applies within the analytical process and in no way constitutes an exportable general truth. To transform the psychoanalyst into an expert in political matters is to ask him to forget this singularity of his knowledge. But this "forgetting" allows him to claim a coincidence between his field of expertise and the problem posed to the community, that is, to redefine drugs in terms that eliminate anything that does not interest the psychoanalyst. And such a redefinition is quite interesting. From now on, the intrinsic harmfulness of some products is no longer an argument; what counts is the fact that it is prohibited and that there is an infraction against this prohibition.

Correlatively, there has been a proliferation of texts of which it is no longer possible to know whether they were written by psychologists, psychoanalysts, or jurists. Discussing the secondary effects of French judicial measures concerning drug addiction, Dr. Henri Guillet passes blithely from the law of 1970 to the law "in general, which regulates the relations of one subject to another and which consequently structures the individual in society," and he turns this assimilation against argument 3 of the Pelletier Report, according to which there is no need to "depenalize" the consumption of "H" since, in reality, the consumer is not the object of severe repression. For Guillet, if the drug addict is searching for the father and the law that the father represents, when he encounters this law it must solidly structure him in a coherent manner and engage him in a "necessary work of symbolization." Consequently, to not firmly prosecute those who violate the prohibition is to find oneself "in a situation where the law neither fulfills its social function of delimiting a space, nor its paternal function of nomination and symbolization for the individual."[11] What a curious drift within a democratic country to confer upon the state a "paternal" role that Lacanian psychoanalysis, moreover, refuses to attribute to any particular individual since it precisely draws part of its fascinating power by not confusing the Father and daddy, the Phallus and the penis, the instance of the Superego and the different authorities embodied in this world. A curious drift, correlatively, of the notion of penal law, when one demands that it fulfill the function of imposing prohibitions when this is not its function. Because, as Cartuyvels and Kaminski remind us,

the penal law does not prohibit; it formulates the possibility of the act in its hypothesis and provides it with a sanctioning apparatus. The penal law is positive in two senses: it is positive because it constitutes a historical event whose appearance falls within the political domain, that is, the provisional solution of a conflict of interests. . . . But the penal law is also positive in that it contains no negation: the law that punishes the drug user does not prohibit (or prevent) anything. It is a matter of a tariff, indicating the price of the drug. The penal law no more entails "no drugs" than any other use of the code, to the despair of the mythologists.[12]

The circle is completed when a magistrate, Dominique Charvet, former president of MILT (Interministerial Mission on the Fight against Drug Addiction), assimilates perfectly the consensual message. Drug addicts distancing themselves, opposing an "elsewhere" to the common world, must be called on to liberate themselves from their enslavement in the name of the "democratic wager of clarity and liberty."[13] Charvet discusses the virtues of a transubstantiation of the "judicial dialogue," which, based as it was on the principles of the autonomy of will and the inalienability of freedom, now becomes "an enunciation bearing on these values," and of the judge who, far from washing his hands by sending the drug addict who accepts the therapeutic injunction to those who will "cure" him, must open up a dialogue with him that calls him to responsibility, to the participation in the "human project based on belief and action with a view to the liberation of physical and moral constraints." If one pursues this line of reasoning, the following conclusion is inevitable: it is not really the users of illegal drugs, but all those who "distance themselves," thanks to any product (or without any product), who should meet the judge. When therapists accept the confusion between juridical law and symbolic law, they create judges who, in turn, intend, in the name of the state, to delve into hearts and souls, transforming themselves into "civil therapists."

Such are the miracles of moral consensus. The premise "don't take drugs" led first to psychologize the law then to "juridicize" the psychological problem. All the cats have turned grey and one no longer knows exactly what one is talking about—the problem posed by drug addiction, the pertinence of a law, the hardly democratic relation between the individual and the law when it is stated in psychoanalytical terms, or the foundations of a democratic society. The use of drugs, infraction of the law, already "signifies" drug addiction, a desperate message to an absent law. The categorization between legal and illegal has become the manifestation of the "arbitrariness of the signifier," which is indifferent to reality since it is this that structures reality. And the law has finally become a project requiring not respect but adherence. "You must be free," says the judge. A paradoxical injunction to subjugation to autonomy.

T W E L V E

Body Fluids[1]

(with Didier Gille)

NOT ONLY is it blood and sperm, but probably saliva, and perhaps even mother's milk. The infectious agent responsible for AIDS gradually reveals the diversity and the unity of its modes of transmission: generally referred to as "body fluids," they are those fluids that (contrary to extremists) are an intrinsic production of the body and can, in orientation or by accident of design, both leave and enter it.

In any case, that is what the media informs us. Some specialists uphold the contrary view that, as with the hepatitis virus, blood is solely responsible; and the current debate has taught us there are many discreet opportunities for mixing our blood. Be that as it may, the discovery that AIDS can be transmitted in a variety of ways increased considerably the potential spread of the disease. We know that AIDS is an epidemic. Yet today the epidemic is no longer exclusive to those "risk groups" so designated by public morality: homosexuals and drug users. They are unique simply because they were among the first to be stricken, unique in a purely quantitative sense, linked to probability, to frequency. Any of us could be affected.

This increase in the spread of the epidemic, which makes AIDS commonplace and the basis of a whole medical industry, does not take it out of the domain of myths, quite the contrary. Having become a disease epidemic in others, it revives the haunting memory of the very phenomenon that is unique to the epidemic. Contrary to functional diseases, epidemic diseases seem to strike arbitrarily: there is no direct causal link between the individual's way of life, the care he takes or does not take of his body, and his exposure to the disease; there is no absolute similarity between the mode of transmission of the disease and its symptoms. The fear of epidemic is the abstract fear of relations between one and many, of the endless multiplication of the one. One virus and one person are enough to cause an outbreak. One contact and one person are enough, directly or indirectly, with any member of an undefined and anonymous group of carriers, and not only are you instantly stricken, but you change sides, become a member of the menacing hoard, able in turn to pass on and advance the disease. Who is the enemy? It could be anyone and, overnight, even oneself.

In *Les Microbes*: guerre et paix; Irréductions Bruno Latour has described this consequence of the Pasteurian "revolution": we are not the numbers we are led to believe.

> *There are not only "social" relations, relations between man and man. Society is not made up just of men, for everywhere microbes intervene and act. We are in the the presence not just of an Eskimo and an anthropologist, a father and his child, a midwife and her client, a prostitute and her client, a pilgrim and his God, not forgetting Mohammed his prophet. In all these relations, these one-on-one confrontations, these duels, these contracts, other agents are present, acting, exchanging their contracts, imposing their aims, and redefining the social bond in a different way. Cholera is no respecter of Mecca, but it enters the intestine of the hadji; the gas bacillus has nothing against the woman in childbirth, but it requires that she die. In the midst of so-called "social" relations, they both form alliances that complicate those relations in a terrible way.[2]*

The "microbes" blur our distinctions, preventing the identification of the person intrinsically responsible, of an intrinsically dangerous relation or situation. Syphilis could function as a sign of the curse on sterile relationships, between man and prostitute. And, so far as the theory of miasmas prevailed, which saw cholera as the translation into bodies of the squalid living conditions of the poor, it could function as a pathogenic indicator of the social question of poverty. But if, as the Pasteurians proposed, cholera is not born of determinable circumstances, if its propagation is quite indifferent to class and way of life, how can it be prevented? The panic behind this question is echoed in the *Belgian Bulletin of the Red Cross* in 1873:

> *By admitting the spread of disease from one country to another, we really accept our powerlessness. How do you catch a demon that is everywhere, from the molecules of the winds and waters*

to the evacuations of the ill? How do you distinguish, in the midst of this whirlpool, of these waves of human diversity that everywhere inundate the globe, and France in particular, between those infected with cholera and those who are not, and stop them before they can spread the plague germs? [But, on the contrary], if cholera is a spontaneous disease, arising in us from a conjunction of determinable circumstances, from a number of knowable causes, the doctors and health authorities will only have to turn their minds to understanding the conditions responsible for this "black plague," and science will deliver us from it.

In some respects, the present situation is the direct opposite to that denounced by the health authorities hostile to the Pasteurian revolution. Many fear that AIDS will revive the denunciation of homosexual practices as abnormal, condemned by nature. From this point of view, it is a relief to find that AIDS is no longer "the homosexual disease," as cholera was "the disease of slum areas," that the latest books and articles announce to everyone, "it concerns *you.*" Whatever the suffering, at least we are spared the hand of God, or of nature, that strikes down the deviant; at least we come back to that abstract fear of the epidemic.

In this regard, the growing medicalization of the problem of AIDS is a reassuring process: the curse gives way to the illness. The specter departs, we can breathe. But words have been uttered about "risk groups," and particularly homosexuals, words that continue to murmur away and sometimes break to the surface. Ponderous words, which now have to be pondered.

The racist blunder that wanted to link homosexual practice to AIDS must not blind us to the intention of other, more frequent and better accepted words, those that have stigmatized, beyond homosexuality, the idea of sexual liberation: the practice of multiple sexual relations, in which certain homosexuals have become the experimenters. And that, as an established fact: the regrettable conclusion was that this liberation, these multiple relations, constituted in themselves a transgression against the order of nature.

Transgression against "the order of nature"? But which order? For nature today no longer offers the comforting image of law and order, of stability and harmony, but of something quite dangerous and proliferating. In *Plagues and Peoples*, William McNeill has shown how the explorations, conquests, commercial ventures, and changes in lifestyle punctuating the history of mankind were accompanied, like some clandestine understudy, by the history of epidemics.[3] Whenever segments of unrelated populations come into contact, whenever the exchanges between men are transformed, infectious agents find the occasion for new malignancies. When it concerns bacteria and viruses, nature does not have the obliging or balanced face that ecologists ascribe to our scruples. It is ruthlessly opportunistic. New paths, new modes of attack are endlessly invented in our person, as in the person of any other living being. To each new vaccine or antibiotic of human invention there corresponds the invention of a new virulent form immune to this defense.

Today we have eliminated many paths of infection and controlled many others. The water we drink is disinfected, food is controlled, suspect animals are shot, and our spit is deflected by the transparent cages that protect officials. There are not many contacts with nature left that are sufficiently unmonitored or common enough to enable the epidemic agents to distort and reinvent their meaning. That only leaves the exchange of body fluids. And we can be sure that AIDS will not be the last disease to infect us in this way, which up to now has received little attention. A disease like AIDS does not represent a transgression with regard to "the order of nature," but profits from the last contacts to escape control, those contacts whose meaning can still be exploited by the "natural" beings with which we coexist.

Sexual liberation is presented far too often as the triumph of culture over nature. A feeble triumph, in truth, since in the human species males are not restricted to mating, nor women to the estrous cycle of childbearing. That is how things are. One might want to have sex all the time, with anyone. It is natural, it is written in our chromosomes. But, if human culture invented anything, it was precisely the means of channeling nature, of replacing the constraints of biology by those of institutions. In that sense, sexual liberation does not really turn its back on nature, but on culture.

This is where you have to proceed with caution. The terrain is mined with pathos, with commonplaces and good intentions, because what this liberation seemed to turn its back on really functioned to protect us from a dangerous nature. Yes, the exchange of body fluids is dangerous; it is dangerous like life itself, which does not move in a closed circle, but as an endless flowing in and out of things. The living being is an open system, open to an environment that is not only nurturing, but peopled with other living beings pursuing their own ends. So must we "return to culture," applaud the restrictions, norms, and usages that channel the flows and protect us?

Some people today are faced with having to make a decision about their way of life. Risk, responsibility, and sexual appetite pose a wide range of problems. The outcome of such decisions is not our concern, but rather the manner in which they are being described.

One small remark to begin with, obvious but interesting. AIDS did not originally strike just homosexuals and drug users, but certain African populations and Haitians as well. But the deaths would have gone on for ages in Zaire and Haiti before anyone thought to notice it or do something about it, because people are dying there all the time. It is not the same for whites, even homosexuals. In this sense, it could be argued that, by undoubtedly contributing to the spread of the virus, homosexuals have at the same time alerted us to its existence. And the existence of a new and deadly virus is an event that concerns not only a certain segment of the population, but everyone. We believe that it is important to stress this aspect of the question to those cold-blooded people who small-mindedly translate "risk groups" as "groups posing a risk to them." The drunk who drives down the wrong side of the road creates a danger that would not have existed without him. Gays were the discoverers of AIDS, but not its cause. On the contrary, the so-called risk groups are in a sense "advance scouts," the first to be stricken by a danger threatening everyone, but also who can report it and alert others to it.

This is, of course, a retrospective description; the "risk groups" have not wanted to play this role, but they have objectively taken it up. But now that the risk of AIDS is known and it seems likely that other little nasties will take the royal road of our body fluids, what will we say to those who ignore advice and continue to make contacts known to be at risk? Will we treat them as irresponsible, to be lectured to, put under observation, and converted? In that case, our future scenario is assured: that of the child in the glass bubble, for whom the outside environment means death; that of the obsessional struggle against all unmonitored contact as potentially the source of death. Science fiction has already drawn the ultimate conclusions from this scenario, as one reads, for example, in Isaac Asimov's *Caves of Steel* or *The Sun Shines Bright*.

Will we take their role as advance guard seriously, and justify their individual choice in the name of the collective good? A new morality is possible, for all you moralists out there. No worse than any other, and just as idiotic. Free love? A final proof of good citizenship. The old campaigners for sex will sing their epidemic tales around the fire at night, monuments will be erected, and young volunteers will be dispatched to the libidinal front, congratulated by an attentive and grateful medical corps, acclaimed by a captivated and fear-ridden humanity.

Or will we recognize in them a modern form of hero? Not the military hero, not the volunteer for a suicide mission, not the poor sucker put in the moral position of having to "choose" between sacrifice and cowardice. No, the utter fool, the one who, in his life or in his death, does not want to serve as any model, but who accepts grave risks in the name of something that defines his uniqueness, but also exceeds it, and so cannot be shared, but only recognized by others. The hero whose only reward, medal, or crowning glory for the risks he takes is the simple recognition of the exigencies that drive him. It is not our place to define these exigencies, this "something." But we can recognize, down through the ages, the subversive insistence of this question of those who agree to expose their body to danger, not in the name of a country, religion, or conviction, but for an abstract, faceless idea—perhaps what Sade called "Nature."

What these heroes can teach us is infinitely more precious than the self-denial or unconscious of recorded heroes: they explore in their flesh, for pleasure or from passion, what a body is, what it can and cannot tolerate. They tell us and remind us what we are—in this case, producers and consumers of body fluids. Living beings, in danger of life.

Translated by Paul Foss

Notes

Chapter 1. Complexity: A Fad?

1. Hubert Reeves, *Atoms of Silence: An Exploration of Cosmic Evolution*, trans. Ruth A. Lewis and John S. Lewis (Cambridge: MIT Press, 1984).

2. Jean-Marc Lévy-Leblond, *L'Esprit de sel* (Paris: Fayard, 1981).

3. Pierre Berge, Yves Pomeau, and Christian Vidal, *Order within Chaos: Towards a Deterministic Approach to Turbulence*, trans. Laurette Tuckerman (New York: John Wiley and Sons, 1986).

4. P. Grassberger and I. Procaccia, *Physica. D*, vol. 9 (1983): 189–208.

5. On this point, see Ilya Prigogine and Isabelle Stengers, *Order out of Chaos* (New York: Bantam, 1984).

6. See Ivor Ekeland, *Mathematics and the Unexpected* (Chicago: University of Chicago Press, 1988).

7. Henri Atlan, *Entre le cristal et la fumée* (Paris: Éditions du Seuil, 1979).

8. This is the position defended in *Order out of Chaos*.

9. Jacques Monod, *Chance and Necessity* (New York: Vintage Books, 1972).

10. Conrad Waddington, *The Strategy of the Genes* (London: Allen and Unwin, 1957).

11. See, notably, Stephen Jay Gould, *Ontogeny and Phylogeny* (Cambridge: Belknap Press of Harvard University, 1977).

12. Judith Schlanger, *Penser la bouche pleine* (Paris: Fayard, 1983).

13. Thomas S. Kuhn, *The Structure of Scientific Revolutions* (Chicago: University of Chicago Press, 1970).

Chapter 3. The Reenchantment of the World

1. Steven Brush, "Irreversibility and Indeterminism: From Fourier to Heisenberg," *Journal of the History of Ideas*, vol. 37 (1976): 603–30.

2. Serge Moscovici, "Quelle unité de l'homme?" in *Hommes domestiques et hommes sauvages* (Paris: Christian Bourgois, 1974), pp. 10–18.

3. Described in Henri Bergson, *Mélanges* (Paris: PUF, 1972), pp. 1340–46.

4. Albert Einstein and Michele Besso, Correspondence 1903–1955, ed. P. Speziali (Paris: Hermann, 1972).

5. Gérard Granel, "Husserl," in the *Encyclopedia Universalis* (Paris, 1971); reprinted with Husserl's "La crise de l'humanité européene et la philosophie" by Paulet publications (Paris, 1975). This small volume shows that the idea of the West's "human mission" is not the sole prerogative of scientists.

6. Maurice Merleau-Ponty, *Résumé de cours 1952–1960* (Paris: Gallimard, 1968), p. 119.

7. Maurice Merleau-Ponty, "Le philosophie et la sociologie," in *Éloge de la philosophie*, Collection Idées (Paris: Gallimard, 1960), pp. 135–37.

8. For what follows, see also Ilya Prigogine, Isabelle Stengers, and S. Pahaut, "Dynamics from Leibniz to Lucretius," Afterword to Michel Serres, *Hermes: Literature Science Philosophy* (Baltimore: Johns Hopkins University Press, 1982), 137–55.

9. On this subject, see Ivor Lerclerc, *Whitehead's Metaphysics* (Bloomington: Indiana University Press, 1975).

10. These perspectives are developed in Ilya Prigogine, *From Being to Becoming* (San Francisco: W. H. Freeman, 1980).

11. Alfred North Whitehead, *Process and Reality* (New York: Free Press, 1978), p. 208.

12. Michel Serres, *La Naissance de la physique dans le texte de Lucrèce* (Paris: Éditions de Minuit, 1977), p. 139.

13. Lucretius, *De rerum natura*, book 2, ed. and commentary C. Bailey, 3 vols. (Oxford: Oxford University Press, 1947). "Again, if all movement is always interconnected, the new arising from the old in a determinate order — if the atoms never swerve so far as to originate some new movement that will snap the bonds of fate, the everlasting sequence of cause and effect — what is the source of the free will possessed by living things throughout the earth?"

14. Serres, *La Naissance de la physique dans le texte de Lucrèce*, p. 136.

15. Ibid., p. 162.

16. Thomas S. Kuhn, *The Structure of Scientific Revolutions*, 2d enlarged ed. (Chicago: University of Chicago Press, 1970).

17. Maurice Merleau-Ponty, *Résumés de cours 1952–1960* (Paris: Gallimard, 1968), pp. 117–18.

18. Whitehead, *Process and Reality*, p. 17.

19. Gilles Deleuze, *Difference and Repetition*, trans. Paul Patton (New York: Columbia University Press, 1994), pp. xx–xxi.

20. Ibid., p. 220; translation modified.

21. Ibid., p. 219.

22. Ibid., p. 218.

23. Serres, *La Naissance de la physique dans le texte de Lucrèce*, pp. 85–86, and "Roumain et Faulkner traduisent l'Écriture," in *La Traduction* (Paris: Éditions de Minuit, 1974).

24. André Neher, "Vision du temps et de l'histoire dans la culture juive," in *Les cultures et le temps* (Paris: Payot, 1975), p. 179.

Chapter 4. Turtles All the Way Down

1. The proceedings of the Cerisy symposium were published as *L'Auto-organisation, de la physique au politique*, ed. Paul Dumouchel and Jean-Pierre Dupuy (Paris: Éditions du Seuil, 1983).

2. In Gilles Deleuze and Félix Guattari, *A Thousand Plateaus: Capitalism and Schizophrenia*, trans. Brian Massumi (Minneapolis: University of Minnesota Press, 1987), one finds an opposition between Celeritas and Gravitas that accords with what we are going to establish between, on the one hand, a science of processes, kinetics, which deviates from a difference that does not contradict accounts but makes them irrelevant, and, on the other hand, a science of falling bodies, of engineers and accounts, like those dominated by conservation, the science of "potential" functions: "*Slow and rapid are not quantitative degrees of movement but rather two types of qualified movement*, whatever the speed of the former or the tardiness of the latter. Strictly speaking, it cannot be said that a body that has dropped has a speed, however fast it falls; rather it has an infinitely decreasing slowness in accordance with the law of falling bodies. Laminar movement that striates space, that goes from one point to another, is weighty; but rapidity, celerity, applies only to movement that deviates to the minimum extent and thereafter assumes a vortical motion, occupying a smooth space, actually drawing smooth space itself. In this space, matter-flow can no longer be cut into parallel layers, and the movement no longer allows itself to be hemmed into biunivocal relations between points. In this sense, the role of the qualitative opposition gravity-celerity, heavy-light, slow-rapid is not that of a quantifiable scientific determination but that of a condition that is coextensive to science and that regulates both the separation and the mixing of the two models, their possible interpenetration, the dominion of one by the other, their alternative" (p. 371; emphasis added).

3. Émile Meyerson, *Identité et Réalité* (Paris: Vrin, 1951), pp. 325–26.

Chapter 5. Black Boxes; or, Is Psychoanalysis a Science?

1. Sigmund Freud, *La technique psychanalytique* (T.P.) (Paris: PUF, 1985). (This corresponds in large part to *Papers on Technique*, vol. 12, *The Standard Edition of the Complete Psychological Works of Sigmund Freud*, ed. and trans. James Strachey [London: Hogarth Press, 1958].)

2. Evariste Galois (1811–32), French mathematician remembered for the Galois theory of groups and other contributions to higher algebra, including the Galois imaginaries.

3. On this subject, see Michel Schiff, *The Memory of Water: Homeopathy and the Battle of Ideas in the New Science*, with a foreword by Jacques Benveniste (London: Thorsons, 1995).

4. A collective work published by Seuil in 1987.

5. On this subject, see Bruno Latour and Steve Woolgar, *Laboratory Life: The Construction of Scientific Facts* (Princeton, N.J.: Princeton University Press, 1986).

6. An interesting question for French psychoanalysts: why is it among them that one finds the most determined representatives of the characterization of the "subject of science" as such? In doing this, what agenda are they following?

7. Bruno Latour, *The Pasteurization of France*, trans. Alan Sheridan and John Law (Cambridge: Harvard University Press, 1988) (followed, in the same edition, by the equally indispensable *Irreductions*).

8. See Latour and Woolgar, *Laboratory Life*, as well as Bruno Latour, *Science in Action: How to Follow Scientists and Engineers through Society* (Milton Keynes: Open University Press, 1987).

9. Stephen J. Gould, *Ever since Darwin: Reflections in Natural History* (Harmondsworth, England: Penguin, 1980), *The Panda's Thumb: More Reflections in Natural History* (Harmondsworth, England: Penguin, 1983), *Hen's Teeth and Horses' Toes: Further Reflections in Natural History* (Harmondsworth, England: Penguin, 1984), and *The Flamingo's Smile* (Harmondsworth, England: Penguin, 1991).

10. I am here paraphrasing the critical account of André Green's theses by Jean Laplanche in *Nouveaux fondements pour la psychanalyse* (Paris: PUF, 1987). One could also look with interest at Serge Lebovici, *Le nourrisson, la mère et le psychanalyste. Les interactions précoces* (Paris: Éditions du Centurion, 1983), in which the "psychoanalytic child" defends its claims to unify the different theoretical children, that is, to be itself described as "originary."

11. Robert S. Wallerstein, "One Psychoanalysis or Many?" *International Journal of Psycho-Analysis*, vol. 69 (1988): 5–21.

12. This work has led to Léon Chertok and Isabelle Stengers, *A Critique of Psychoanalytic Reason: Hypnosis as a Scientific Problem from Lavoisier to Lacan*, trans. Martha N. Evans (Stanford, Calif.: Stanford University Press, 1992). See also Léon Chertok and Isabelle Stengers, "Therapy and the Ideal of Chemistry," *Nature*, vol. 329 (1987): 768, and "The Deceptions of Power: Psychoanalysis and Hypnosis," *Substance* 62–63 (1991): 81–91.

13. Sigmund Freud, "Remembering, Repeating and Working-Through," in *The Standard Edition of the Complete Psychological Works of Sigmund Freud*, vol. 12, p. 152; hereafter cited in the notes as *SE*, along with the relevant volume and page numbers.

14. Sigmund Freud, "'Wild' Pschoanalysis," in *SE*, vol. 11.

15. Freud, "Remembering, Repeating and Working-Through," pp. 154–55.

16. Sigmund Freud, "The Dynamics of Transference," in *SE*, vol. 12, p. 106.

17. Sigmund Freud, "Analytic Therapy," in *SE*, vol. 16, p. 451.

18. Sigmund Freud, "Analysis Terminable and Interminable," in *SE*, vol. 23.

19. Laplanche, *Nouveaux fondements pour la psychanalyse*.

20. This is Léon Chertok's basic thesis in *Le non-savoir des psy* (Paris: Payot, 1979).

Chapter 6. Of Paradigms and Puzzles

1. *Of Paradigms and Puzzles* is the transcript of a lecture given by the author in June 1985 at the Limites-frontières Seminar in Paris.

2. Judith Schlanger, *Penser la bouche pleine* (Paris: Fayard, 1983), and *L'Invention intellectuelle* (Paris: Fayard, 1983).

Chapter 7. Is There a Women's Science?

1. Text published as the foreword to the French edition of Evelyn Fox Keller, *A Feeling for the Organism: The Life and Work of Barbara McClintock* (San Francisco: W. H. Freeman, 1983).

2. Albert Dalcq, *L'œuf et son dynamisme organisateur* (Paris: Albin Michel, 1941), p. 541.

3. Robert Musil, *The Man without Qualities* (London: Picador, 1995).

Chapter 8. The Thousand and One Sexes of Science

1. This essay was first presented at a conference in Moncton, Canada, October 1990, on the topic "Les théories scientifiques ont-elles un sexe?"

Chapter 9. Who Is the Author?

1. Quoted in Pierre Duhem, *Sozein ta phainomena. Essai sur la notion de théorie physique de Platon à Galilée* (Paris: Vrin, 1982), p. 134.

2. See notably Éric Alliez, *Capital Times: Tales from the Conquest of Time*, trans. Georges Van Den Abbeele, Foreword by Gilles Deleuze (Minneapolis: University of Minnesota Press, 1995).

3. Bruno Latour, "D'où viennent les microbes," *Les Cahiers de Science et Vie. Les grandes controverses scientifiques n° 4, Pasteur. La tumultueuse naissance de la biologie moderne* (August 1991): 47.

4. In relation to all of this, see Bruno Latour, *Science in Action: How to Follow Scientists and Engineers through Society* (Milton Keynes: Open University Press, 1987).

5. Jean-Pierre Changeux, *Neuronal Man: The Biology of Mind* (New York: Oxford University Press, 1986).

6. Stephen J. Gould, *Wonderful Life: The Burgess Shale and the Nature of History* (London: Hutchinson Radius, 1990).

7. The title of a book by Stephen J. Gould, *The Panda's Thumb: More Reflections in Natural History* (Harmondsworth, England: Penguin, 1983). See also Stephen J. Gould, *The Flamingo's Smile* (Harmondsworth, England: Penguin, 1991), and *Hen's Teeth and Horses' Toes: Further Reflections in Natural History* (Harmondsworth, England: Penguin, 1984).

8. Shirley Strum, *Almost Human: A Journey into the World of Baboons* (New York: Random House, 1987).

Chapter 10. Time and Representation

1. *Bulletin communal de la Ville de Bruxelles*, 1852, p. 54. (These council bulletins are designated by the initials B.C.)

2. This type of transmission was invented in England. See Derek Howse, *Greenwich Time* (Oxford: Oxford University Press, 1980), chapter 4.

3. Starting in 1856, an electrical signal coming from Greenwich Observatory was transmitted to private houses that had a subscription to this service. (See Howse, ibid.)

4. B.C., 1856, p. 523.

5. See Guillaume Bigourdan, *L'Astronomie — Évolution des idées et des méthodes* (Paris: Flammarion, 1911), pp. 78–83.

6. Houzeau, cited in B.C., 1892, p. 411.

7. B.C., 1892, p. 398.

8. Ibid., pp. 404–5.

9. Ibid., p. 402.

10. For a general history of what follows, refer to Howse, *Greenwich Time.*

11. Howse (ibid.) shows how the progress of watchmaking, but also of astronomy (and thus of mathematics) during the eighteenth century is tied to the search for a reliable way of calculating longitudes financed by the state and commercial powers.

12. The cycloids appear as two strips of a mathematically determined form, fixed in the pendulum's plane of oscillation and against which the wire will alternately press (see figure 5). The deformation that is thereby obtained modifies in a continuous way the actual length of the pendulum. This operation is a veritable mathematical transformation, which does not influence the movement of the pendulum or the mechanism,

restricting itself to making the length of the pendulum a function of the angle of oscillation.

13. The pirouette (see figure 3, OP), a supplementary gearwheel, in this case a mechanical procedure, will solve the problem by reducing the amplitude of the pendulum's oscillations, so as to limit them to the domain of values wherein the law of isochronism is approximately valid. The abandonment of the cycloids that achieved the exact mathematical production of the law of isochrony to the profit of a mechanism that approaches this law by gearing down, shocks, and friction marks the passage to a more operational conception of the production of a lawful phenomenon.

14. Edmond About, *Dans les ruines* (1867), *Recherches*, no. 25 (November 1976): 77.

15. Louis Reybaud, *Le Fer et la Houille* (1874), *Recherches*, no. 25 (November 1976): 118.

16. B.C., 1854, pp. 198–203.

17. The opposition irregular foliot/isochronal pendulum clock is false. The physical law that controls the course of the foliot (the law of inertia) implies, in ideal conditions that eliminate friction, an operation just as regular as that of a pendulum in the same conditions. The great difference is that in the one case the collisions, which are always difficult to control, only put into play a marginal quantity of energy, whereas in the other, they are responsible with each blow for the totality of contributions and losses of energy.

18. See D. Caldwell, *From Watts to Clausius* (London: Heinemann, 1971).

19. George Cournol, *Considérations d'intérêt public sur le droit d'exploiter les mines* (Paris: Académie-Française, 1790).

20. Edouard Ducpetiaux, *Colonies agricoles, écoles rurales et écoles de réforme pour les indigents…*, 1851, p. 19.

21. *Enquête de 1843*, vol. 2, p. 370, in Jean Neuville, *La Condition ouvrière au XIXᵉ siècle*, vol. 1 (Brussels: Vie Ouvrière, 1976), p. 109.

22. Scrive Frères, *Note sur la situation des ouvriers de l'établissement de tissage mécanique*, 1851, *Recherches*, no. 25 (November 1976): p. 114.

23. Joseph Billig, *Les Camps de concentration dans l'économie du Reich hitlérien* (Paris: PUF, 1973), pp. 263–64.

Chapter 11. Drugs: Ethical Choice or Moral Consensus

1. Antoine Lazarus, "Tous Prévenus," in *L'Esprit des Drogues, Autrement*, April 1989, pp. 94–95.

2. Marie Andrée Bertrand, "L'immoralité de la prohibition," *Psychotropes*, vol. 5, nos. 1–2 (1989): 17.

Further on, we will see that the Pelletier Committee's report, quoted by the author, presents as an argument in its defense of the nondepenalization of the use of cannabis the fact that cannabis is not "as severely punished as some people claim." The argument, however, is not accompanied by any statistics that would, for example, allow one to respond to more specific "claims," bearing, for example, on the ethnic origin of the imprisoned users. In fact, given that all drug offenses are lumped together under French law, it is, even today, difficult to know what percentage of imprisonments is for cannabis use, or on what basis it is decided to prosecute, since the initiative is given to the police and the public prosecutor's department. Nevertheless, in the Trautmann Report, one reads that, in 1988, 60.3 percent of police interrogations on the use of narcotics and psychotropic drugs were concerned with cannabis. The report does not specify the proportion of these interrogations that led to prosecutions.

3. The law of December 31, 1970, provides for exceptionally severe procedures and penalties with respect to the common law and includes excessive provisions both at the level of police custody and police searches. As Francis Caballero emphasizes in *Droit de la Drogue* (Paris: Dalloz, 1989), notably pp. 589–91, the legislation in force defines a real law of exception, where a correctional procedure is used with the backup of criminal sentencing as a key element.

4. Ibid., pp. 489–91.

5. Thus, in our preceding citation from the Trautmann Report, the abolitionists are accused of refusing to "listen to the message," although they have, of course, never suggested that psychological assistance be denied to drug users and addicts; this "listening" is thus a function attributed to the penal code by the report.

6. The Trautmann Report, p. 109.

7. Ibid., p. 205.

8. The Trautmann Report devotes considerable attention to the theme of "drug addiction and adolescence" and focuses the obligations of the state on

this age group: "Can one imagine that the state, being responsible for the future of the country, and in particular for its youth, would cold-bloodedly participate in the poisoning and destruction of them?" (p. 207). A few pages earlier (p. 197), the same report assumes, quite curiously, that the use of drugs by minors constitutes a thorny problem for the abolitionists. Apparently, the existence of legislation limiting the access of minors to a legal drug like alcohol has been forgotten.

9. Michèle Barzach, *Le paravent des égoïsmes* (Paris: Éditions Odile Jacob, 1989), pp. 64–67.

10. Forgetting—as Yves Cartuyvels and Dan Kaminski remind us in "Dépénalisation des drogues: articulation socio-politique et clinique," a contribution to the Ninth Days of Reims for a Clinic on the Drug Addict (December 1–2, 1990)—that in good Lacanian doctrine the symbolic law can only be transgressed in a phantasm, which leads them to ask the question: "And if the difference—the 'little difference'—between the drug user and others only amounted to this: that others think that the drug user has succeeded. A hypothesis: only the other of the drug user thinks that the latter has transgressed."

11. *Acts of the Fifteenth International Conference on the Prevention and Treatment of Drug Dependency* (1986), p. 277.

12. Cartuyvels and Kaminski, "Dépénalisation des drogues."

13. Dominique Charvet, "La Justice aux prises avec l'intime," *Autrement*, April 1989, pp. 64–69.

Chapter 12. Body Fluids

1. Chapter 12 originally appeared in *L'Autre Journal* 10 (December 1985): 110–20. The original translation has been modified.

2. Bruno Latour, *The Pasteurization of France*, trans. Alan Sheridan and John Law (Cambridge: Harvard University Press, 1988), p. 35.

3. William McNeill, *Plagues and Peoples* (New York: Doubleday, 1977).

Index

Isabelle Stengers is an associate professor of philosophy at the Free University of Brussels and a Distinguished Member of the National Committee of Logic and the History of Philosophy of Sciences in Belgium. She was the 1993 recipient of the Grand Prix in philosophy from the French Academy and is the author of numerous books and articles on the history of science, the theory of systems, and psycho-analysis. Among them are *The Will to Science: About Psychoanalysis*, *A History of Chemistry* (with Bernadette Bensaude-Vincent), and, with Ilya Prigogine, *Order out of Chaos*.

Bruno Latour is a professor at the Centre de Sociologie de l'Innovation, École Nationale Supérieure de Mines de Paris. He is the author of *Science in Action*, *The Pasteurization of France*, *We Have Never Been Modern*, and *Aramis, or the Love of Technology*, among other works.

Paul Bains is a research scholar in The School of Humanities, Murdoch University, Western Australia, and co-translator of Félix Guattari's last work, *Chaosmosis: An Ethico-Aesthetic Paradigm*.